GCE A-Level

Pure Mathematics

R. A. Parsons, B.Sc.
and
A. G. Dawson, B.Sc.

Published by Charles Letts Books Ltd
London, Edinburgh and New York

This book is sold subject to the condition that it shall not, by way of trade or otherwise, be lent, re-sold, hired out, or otherwise circulated without the publisher's prior consent in any form of binding or cover other than that in which it is published and without a similar condition including this condition being imposed on the subsequent purchaser.

Published 1983 by Charles Letts Books Ltd
Diary House, Borough Road, London SE1 1DW

1st edition   1st impression
© Charles Letts Books Ltd
Made and printed by Charles Letts (Scotland) Ltd
ISBN 0 85097 558 1

ROD PARSONS graduated from Bristol University and completed a post-graduate Certificate in Education at Southampton University. Since then he has taught in schools and sixth form colleges in Surrey. He has taught both modern and traditional syllabuses, has written for S.M.P. and lectured to teachers at S.M.P. conferences. He was Head of the Mathematics Department at Sunbury Sixth Form College and after that Head of Mathematics at Woking Sixth Form College. He is now an Assistant Principal at Esher Sixth Form College.

TONY DAWSON graduated from Southampton University with an honours degree in Mathematics. After completing his post-graduate Certificate in Education, also at Southampton University, he began teaching mathematics at Wallington Grammar School for Boys. Since then he has taught the subject to all levels of ability including Oxbridge Entrance and his teaching has covered both modern and traditional syllabuses. He has had considerable experience as a Head of Department at Woking Grammar School for Boys. More recently he has been appointed Deputy Principal of Woking Sixth Form College, but is still actively involved in preparing candidates for advanced level examinations.

Worked Examples

APPLIED MATHEMATICS, E. M. Peet, B.Sc.

BIOLOGY, A. G. Toole, B.Sc.

CHEMISTRY, J. E. Chandler, B.Sc.
and R. J. Wilkinson, Ph.D.

ECONOMICS, R. Maile, B.A.

GEOGRAPHY, C. Lines, M.Sc.

PHYSICS, G. M. George, B.Sc.

PURE MATHEMATICS, R. A. Parsons, B.Sc.
and A. G. Dawson, B.Sc.

PURE AND APPLIED MATHEMATICS,
R. A. Parsons, B.Sc. and A. G. Dawson, B.Sc.

Contents

Introduction

This book of worked examples is designed as a sequel to *Pure Mathematics* by the same authors in the Key Facts series and is set in the style of Advanced Level Mathematics examination questions.

The chapters follow the same headings as those in the Pure Mathematics book with an additional section on probability. Most examination syllabuses have a common core of content but may vary considerably in the range and depth of the extra topics which they include. The book attempts to illustrate examples of different styles of examination question as appropriate in the various chapters and these have been grouped into the following categories:

Multiple choice questions (Type A) Select the correct answer
In this type of question five possible responses are given and the candidate is required to decide which one is correct.

Multiple choice questions (Type B) Answer according to the table

A	B	C	D	E
1, 2, 3	1, 3	2, 3	2	3
correct	only	only	only	only

Three statements are given, each of which may be true or false. Each statement should be considered separately and the final choice made according to the above table.

Multiple choice questions (Type C) For each question two statements are given. Answer

A if 1 implies 2 and 2 implies 1
B if 2 implies 1 but 1 does not imply 2
C if 1 implies 2 but 2 does not imply 1
D if 1 denies 2 and 2 denies 1
E if none of these relations holds

Students often find this type of question difficult but it does promote a good understanding of mathematical concepts. Note that solution A suggests 1 is a necessary and sufficient condition for 2 but solution B implies that 1 is only a necessary condition for 2.

Multiple choice questions (Type D) Each question consists of a problem followed by four pieces of information. Decide whether the problem can be solved with one of the four pieces of information omitted and answer

A if 1 could be omitted
B if 2 could be omitted
C if 3 could be omitted
D if 4 could be omitted
E if none can be omitted

Short questions Some Boards set two or three papers and these may consist of a section A containing 8 to 12 short questions followed by a choice of longer questions or alternatively a whole paper may involve shorter type problems. These usually test an application of just one topic or idea.

Long questions These examples illustrate the type of question the candidate has to answer when faced with a choice of four out of six or seven out of eleven. These questions generally contain two or three applications on one theme or may require processes from different branches of the subject.

The examples we have presented are designed to cover most of the topics examined on an A-Level syllabus for Pure Mathematics. These are based on our experience of teaching Advanced Level Mathematics over many years while preparing students for different examination Boards.

As teachers we recognize that we are leading students to understand and appreciate mathematics more fully and not just preparing them for examinations. Mathematics is an art and possesses a beauty of its own which gives satisfaction to those who master it. However, it is also a tool required in many other areas and subjects and teachers will attempt to adequately cover both of these aspects.

Mathematics is, by its nature, a logical subject and we would encourage students to set out their answers in a clear and coherent manner. These examples will give some idea of the way in which solutions should be presented.

Although we feel that coaching specifically for examinations is not educationally very sound there can be no doubt that 'practice makes perfect' and a worked example may be a distinct aid to learning.

We would advise students to use this book constructively and would suggest that they try to answer the questions before referring to our explanations. Sometimes, however, some guidance may be necessary from the outset and, if this is the case, the student should read the first few lines of the solution and then try to carry the rest of the argument through for himself.

It may well be helpful to cover the working with a card and only reveal each successive line of explanation after trying that stage for oneself.

Mathematics is a subject that cannot easily be mastered by reading alone. It is vital for students to have paper and pencil available so that the steps of the working can actually be written out. Note that, while we have endeavoured to include all the necessary working, students may need to amplify certain stages to fully understand the solutions.

It is good practice to make full use of sketches and diagrams when writing out solutions. For reasons of space we have kept the number of diagrams to a minimum but we would encourage students to draw them for themselves.

Finally, since the style of examination papers and questions varies from Board to Board we would suggest that students spend some time during their examination preparation familiarizing themselves with some of the past examination papers set by their own particular examination board.

Chapter 1

Algebra

Multiple choice questions (Type A) Select the correct answer

Example 1

When $f(x) = x^3 + 2x^2 + 3x + 4$ is divided by $x+2$ the remainder is

A 2 **B** -2 **C** 14 **D** 24 **E** 0

This is a straightforward question which is best answered by using the remainder theorem. The remainder is given by $f(-2)$.

$$\text{Remainder} = f(-2) = (-2)^3 + 2(-2)^2 + 3(-2) + 4$$
$$= -8 + 8 - 6 + 4 = -2 \qquad \text{ANSWER B}$$

Example 2

The sum of the roots of $2x^3 - 3x^2 + 4x - 5 = 0$ is equal to

A -3 **B** 3 **C** $-1\frac{1}{2}$ **D** $1\frac{1}{2}$ **E** $-2\frac{1}{2}$

If the roots (answers) of the equation $ax^3 + bx^2 + cx + d = 0$ are α, β and γ, then

$$\alpha + \beta + \gamma = -\frac{b}{a}; \quad \alpha\beta + \beta\gamma + \alpha\gamma = \frac{c}{a}; \quad \alpha\beta\gamma = -\frac{d}{a}$$

The sum of the roots of the equation is equal to $-\dfrac{(-3)}{2} = 1\frac{1}{2}$.

There is no need to solve the equation completely.   ANSWER D

Example 3

If $\dfrac{5}{x^2 - x - 6} \equiv \dfrac{a}{x+2} + \dfrac{b}{x-3}$ then

A $a = 1$ **B** $b = 1$ **C** $a = 2$ **D** $b = 3$ **E** $a = -2$

The aim of this question is to express $\dfrac{5}{x^2-x-6} = \dfrac{5}{(x+2)(x-3)}$ in partial fractions.

Assuming $\dfrac{5}{(x+2)(x-3)} = \dfrac{a}{x+2} + \dfrac{b}{x-3}$ and multiplying both sides by $(x+2)(x-3)$ gives $5 = a(x-3) + b(x+2)$.

Let $x = 3 \Rightarrow 5 = 5b \Rightarrow b = 1$;   Let $x = -2 \Rightarrow 5 = -5a \Rightarrow a = -1$.

Alternatively, the value of a can be found by substituting $x = -2$ in $\dfrac{5}{(x+2)(x-3)}$ and ignoring the bracket $(x+2)$ whose value is zero.

The value of b is found by putting $x = 3$ and ignoring $(x-3)$.

Therefore, $a = \dfrac{5}{-5} = -1$ and $b = \dfrac{5}{5} = 1$.   **ANSWER B**

Multiple choice questions (Type B) Answer according to the table

A	B	C	D	E
1, 2, 3	1, 3	2, 3	2	3
correct	only	only	only	only

Example 4

A curve is given in parametric form as $x = 2 - 3t$, $y = t + 4$

1 It is a straight line   **2** Its gradient is $\frac{1}{3}$
3 Its intercept on the y-axis is $4\frac{2}{3}$

The best way to answer statement 1 is to try to eliminate t and form the Cartesian (x, y) equation of the curve.

$$x + 3y = 2 - 3t + 3t + 12 = 14 \Rightarrow x + 3y = 14 \text{ (a straight line)}$$

Rearranging the equation gives $y = -\frac{1}{3}x + 4\frac{2}{3}$ which represents a straight line with gradient $-\frac{1}{3}$ and intercept on the y-axis $4\frac{2}{3}$. Statements 1 and 3 are true giving answer B.   **ANSWER B**

Example 5

For the expression $x^3 + y^3$

1 $x+y$ is a factor   **2** $(x+y)^3$ is equivalent
3 $x^2 - xy + y^2$ is a factor

$$x^3 + y^3 = (x+y)(x^2 - xy + y^2)$$
$$(x+y)^3 = x^3 + 3x^2y + 3xy^2 + y^3$$

so statements 1 and 3 are true.
so statement 2 is false.

ANSWER B

Example 6

The region satisfied by the three inequalities contains one point only with integer coordinates.

1 $y > x^2 - 2x, \ x+y < 2, \ y > x$
2 $y > x^2 - 2x, \ x+y < 2, \ y < x$
3 $y > x^2 - 2x, \ x+y > 2, \ y < x$

The diagram in Figure 1 shows the graphs of the three equalities

$$y = x^2 - 2x$$

$$x + y = 2$$

$$y = x.$$

Statement 1 is represented by region Q.

Statement 2 is represented by region P.

Statement 3 is represented by region R.

All three regions contain one integer point only.

ANSWER A

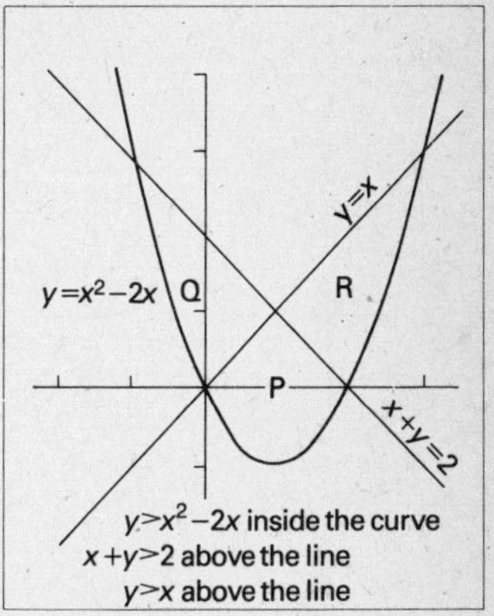

Figure 1

Multiple choice questions (Type C) For each question two statements are given. Answer

A if 1 implies 2 and 2 implies 1
B if 2 implies 1 but 1 does not imply 2
C if 1 implies 2 but 2 does not imply 1
D if 1 denies 2 and 2 denies 1
E if none of these relations holds

Example 7

$$f(x) = ax^2 + bx + c$$

1 $b^2 > 4ac$ **2** $f(x)$ has 2 real and distinct factors

$b^2 > 4ac \Rightarrow ax^2 + bx + c = 0$ has 2 real distinct roots which means that $f(x)$ has 2 real distinct factors. Similarly 2 real distinct factors $\Rightarrow$ 2 real distinct roots $\Rightarrow b^2 > 4ac$. $1 \Rightarrow 2$ and $2 \Rightarrow 1$ which means $1 \Leftrightarrow 2$. **ANSWER A**

Example 8

$$f(x) = ax^4 + bx^3 + cx^2 + dx + e$$

1 $x = 1$ is a point of inflexion   **2** $f''(1) = \left(\dfrac{d^2f}{dx^2}\right)_{\text{at } x = 1} = 0$

At a point of inflexion it is true that $f''(x) = 0$, so $1 \Rightarrow 2$.
If $f''(1) = 0$ it is not necessarily true that $x = 1$ is a point of inflexion. For example, $f(x) = (x-1)^4$ has a minimum point at $x = 1$ but $f''(1) = 0$. So statement 2 does not imply statement 1. **ANSWER C**

Example 9

$a^n + b^n$ (n is a positive integer)

1 n is even   **2** $a + b$ is a factor of $a^n + b^n$

$a + b$ is only a factor of $a^n + b^n$ when n is odd so each statement denies the other. **ANSWER D**

Multiple choice questions (Type D) Each question consists of a problem followed by four pieces of information. Decide whether the problem can be solved with one of the four pieces of information omitted and answer

A if 1 could be omitted   **D** if 4 could be omitted
B if 2 could be omitted   **E** if none can be omitted
C if 3 could be omitted

Example 10

Find the value of $\log_a b + \log_a c - \log_a d$

1 $a = 5$ **2** $b = 2$ **3** $c = 3$ **4** $d = 6$

By the properties of logarithms

$$\log_a b + \log_a c - \log_a d = \log_a\left(\frac{bc}{d}\right) = \log_a 1 = 0$$

whatever the value of a. Therefore, 1 can be omitted. **ANSWER A**

Example 11

Find the limit of $\dfrac{(x+a)(x+b)}{(x+c)(x+d)}$ as x tends to zero.

1 $a = 1$ **2** $b = 2$ **3** $c = 3$ **4** $d = 4$

The limit is $\dfrac{ab}{cd}$ so all four values are needed.   ANSWER E

Short questions

Example 12

(a) Find the set of real values of x for which $x^2 - 9x + 20$ is negative.

(b) Find the set of values of k for which $x^2 + kx + 9$ is positive for all real values of x.

(a) $x^2 - 9x + 20 = (x - 4)(x - 5)$ which takes zero values when $x = 4$ and $x = 5$. The graph of $x^2 - 9x + 20$ is a positive quadratic; the positive x^2 term implies a U-shaped curve with a minimum point. The curve cuts the x-axis at $(4,0)$ and $(5,0)$ and lies below the x-axis for values between these points.

The solution set is $4 < x < 5$ ($x = 4$ and $x = 5$ are excluded).

(b) The graph of $x^2 + kx + 9$ is a positive quadratic (U-shaped) which will be positive for all real values of x if it lies completely above the x-axis. This will happen if there are no real roots to the equation $x^2 + kx + 9 = 0$.

Using the quadratic formula $x = \dfrac{-b \pm \sqrt{b^2 - 4ac}}{2a}$ no real roots occur when $b^2 < 4ac \Rightarrow k^2 < 36 \Rightarrow -6 < k < 6$.

Example 13

Given that $y = ax^n$ draw a suitable straight line graph to determine the constants a and n assuming that the values in the table approximately satisfy the relation (a and n are integers).

x	1	2	3	4	5
y	1.9	16.2	53	130	240

$y = ax^n \Rightarrow \log_{10} y = \log_{10} ax^n$

$\Rightarrow \log_{10} y = \log_{10} a + n \log_{10} x$

Plot $\log_{10} x$ against $\log_{10} y$

$\log_{10} x$	0	0.3	0.48	0.6	0.7
$\log_{10} y$	0.28	1.91	1.72	2.11	2.38

These values are plotted in Figure 2.

The gradient is

$$\frac{2.1}{0.7} = 3 = n$$

The intercept on the y-axis ($\log_{10} a$) is $0.29 \Rightarrow a = 1.95$. Since a is an integer, $a = 2$.

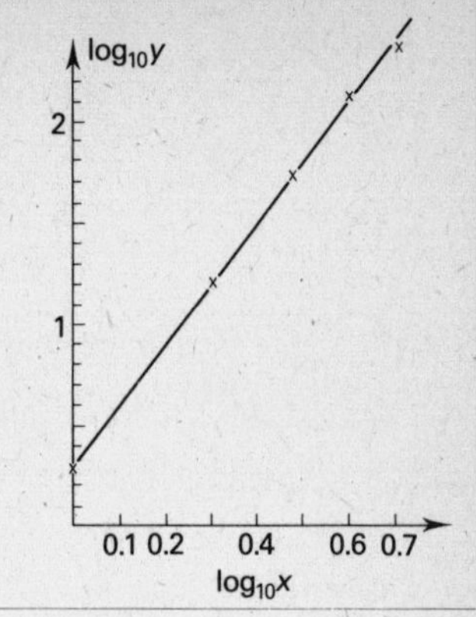

Figure 2

Example 14

If p and q are the roots of $2x^2 + 3x + 4 = 0$ find

(a) $\dfrac{1}{p} + \dfrac{1}{q}$ (b) $p^2 + q^2$ (c) $p^3 + q^3$

(a) It is not necessary to find the exact roots (in fact the roots are not even real).

If p and q are the roots of $ax^2 + bx + c = 0$

then $\quad p + q = -\dfrac{b}{a} = -1\tfrac{1}{2}$ and $pq = \dfrac{c}{a} = \dfrac{4}{2} = 2$

$$\frac{1}{p} + \frac{1}{q} = \frac{p+q}{pq} = \frac{-1\tfrac{1}{2}}{2} = -\frac{3}{4}$$

(b) $p^2 + q^2 = (p+q)^2 - 2pq = (-1\tfrac{1}{2})^2 - 4 = 2\tfrac{1}{4} - 4 = -1\tfrac{3}{4}$

(c) $p^3 + q^3 = (p+q)^3 - 3p^2 q - 3pq^2$

$$= (p+q)^3 - 3pq(p+q)$$

$$= (-1\tfrac{1}{2})^3 - 6(-1\tfrac{1}{2})$$

$$= -\tfrac{27}{8} + 9 = 5\tfrac{5}{8}$$

These questions are usually exercises in algebraic manipulation. They are not difficult but require practice in expressing the quantities required in terms of $p+q$ and pq.

> **Example 15**
>
> Given that $\log_2 x + 2\log_4 y = 4$, show that $xy = 16$. Hence solve for x and y the simultaneous equations
>
> $$\log_{10}(x+y) = 1 \qquad \log_2 x + 2\log_4 y = 4$$

$$\log_4 y = \frac{\log_2 y}{\log_2 4} = \tfrac{1}{2}\log_2 y \text{ using change of logarithm base.}$$

$$\log_2 x + 2\log_4 y = 4 \Rightarrow \log_2 x + \log_2 y = 4 \Rightarrow \log_2 xy = 4$$

This is equivalent to $xy = 2^4 = 16$ (1)

The first equation is equivalent to $x + y = 10^1 = 10$ (2)

Substituting for y from (1) into (2) $x + \dfrac{16}{x} = 10$

$\Rightarrow x^2 + 16 = 10x \Rightarrow x^2 - 10x + 16 = 0 \Rightarrow x = 2$ or $x = 8$ by factors. $x = 2$, $y = 8$ or $x = 8$, $y = 2$ are the two pairs of solutions.

The equations (1) and (2) are symmetrical in x and y leading to the symmetrical solutions.

This type of question requires knowledge of the laws of logarithms:

Basic definition $\log_a b = c \Leftrightarrow a^c = b$ or $2^3 = 8 \Leftrightarrow 3 = \log_2 8$

Change of base $\log_a b = \dfrac{\log_c b}{\log_c a}$, also $\log_a x + \log_a y = \log_a xy$

> **Example 16**
>
> Express in partial fractions $\dfrac{2}{(x+1)^2(x^2+1)}$

$$\frac{2}{(x+1)^2(x^2+1)} = \frac{ax+b}{x^2+1} + \frac{c}{x+1} + \frac{d}{(x+1)^2}$$

These include all the possible factors.

Multiplying both sides by $(x+1)^2(x^2+1)$ gives

$$2 = (ax+b)(x+1)^2 + c(x+1)(x^2+1) + d(x^2+1)$$

$x = -1 \Rightarrow 2 = 2d \Rightarrow d = 1$

Equating coefficients of $x^3 \Rightarrow 0 = a+c$ (1)

Equating coefficients of $x^2 \Rightarrow 0 = 2a+b+c+d$ (2)

Equating coefficients of $x \Rightarrow 0 = a+2b+c$ (3)

Equating constant terms $\Rightarrow 2 = b+c+d$ (4)

$(3) - (1) \Rightarrow b = 0$ and from (4) $2 = 0 + c + 1 \Rightarrow c = 1 \Rightarrow a = -1$

So $\dfrac{2}{(x+1)^2(x^2+1)} = \dfrac{-x}{x^2+1} + \dfrac{1}{x+1} + \dfrac{1}{(x+1)^2}$

Note that the first statement is crucial; a denominator of x^2+1 requires a numerator of $ax+b$ and both parts of a repeated factor are needed.

Example 17
Solve the equation $3^x \cdot 2^{2x} = 7$

Either taking logs to base 10 (although any base will do)

$$x \log 3 + 2x \log 2 = \log 7$$
$$x \, (\log 3 + 2 \log 2) = \log 7$$
$$x \, (\log 3 + \log 4) = \log 7$$
$$x \log 12 = \log 7$$
$$x = \frac{\log 7}{\log 12} = \frac{0.845}{1.079} = 0.783$$

or $2^{2x} = 4^x \Rightarrow 3^x \cdot 4^x = 12^x = 7$

$$\Rightarrow x \log 12 = \log 7 \Rightarrow x = \frac{\log 7}{\log 12} = 0.783$$

Example 18
Given that $\log_{10} 2 = 0.301$ and $\log_{10} 3 = 0.477$ find the values of $\log_{10} 4$, $\log_{10} 5$, $\log_{10} 6$, $\log_{10} 2\frac{2}{3}$ and $\log_{10} 2\frac{7}{9}$.

$\log_{10} 4 = \log_{10} 2^2 = 2 \log_{10} 2 = 2 \times 0.301 = 0.602$
$\log_{10} 5 = \log_{10} \frac{10}{2} = \log_{10} 10 - \log_{10} 2 = 1 - 0.301 = 0.699$
$\log_{10} 6 = \log_{10} 2 + \log_{10} 3 = 0.301 + 0.477 = 0.778$
$\log_{10} 2\frac{2}{3} = \log_{10} \frac{8}{3} = \log_{10} 8 - \log_{10} 3 = 3 \log_{10} 2 - \log_{10} 3$
$\qquad = 0.903 - 0.477 = 0.426$
$\log_{10} 2\frac{7}{9} = \log_{10} \frac{25}{9} = \log_{10} 25 - \log_{10} 9$
$\qquad = 2 \log_{10} 5 - 2 \log_{10} 3 = 1.398 - 0.954 = 0.444$

Example 19
Given that $f(x) = x^2 + bx + c$ has remainder 2 when divided by $x - 1$ and remainder 1 when divided by $x - 2$, find b and c and the remainder when $f(x)$ is divided by $(x-1)(x-2)$.

$$f(1) = 1 + b + c = 2 \Rightarrow b + c = 1$$
$$f(2) = 4 + 2b + c = 1 \Rightarrow 2b + c = -3$$

Thus $b = -4, c = 5 \Rightarrow f(x) = x^2 - 4x + 5$

$f(x) = x^2 - 4x + 5 = (x-1)(x-2) + -x + 3 \Rightarrow$ remainder $= -x + 3$

This question involves the straightforward use of the remainder theorem.

Long questions

> **Example 20**
> If $g(x) = x^3 - 2x^2 - 5x + 6$
>
> (a) solve $g(x) = 0$ (b) sketch the graph of $g(x)$
> (c) find any maximum, minimum or inflexion points

(a) Possible factors of $g(x)$ are $(x \pm 1)$, $(x \pm 2)$, $(x \pm 3)$, $(x \pm 6)$.
Use the factor (remainder) theorem to test for factors.
$g(1) = 1 - 2 - 5 + 6 = 0 \Rightarrow (x-1)$ is a factor.
By inspection (or division) $g(x) = (x-1)(x^2 - x - 6)$
$$= (x-1)(x+2)(x-3)$$
$g(x) = 0 \Rightarrow x = 1$ or $x = -2$ or $x = 3$.

Note that $g(3) = 27 - 18 - 15 + 6 = 0 \Rightarrow (x-3)$ is a factor.
$$g(x) = (x-3)(x^2 + x - 2) = (x-3)(x-1)(x+2) \quad \text{and the}$$
factors are unique no matter in which order you find them.

(b) $g(x)$ is a positive cubic (positive x^3 term $\Rightarrow$ graph goes from bottom left hand corner to top right hand corner).

(i) Look for zeros
$$g(x) = 0 \text{ when}$$
$$x = 1, -2, 3.$$

(ii) $g(0) = 6$ (constant term)

(iii) $g(-1) = -1 - 2 + 5 + 6$
$$= 8$$
$$g(2) = 8 - 8 - 10 + 6$$
$$= -4$$

(iv) Do not assume maximum is at $x = -1$ or minimum is at $x = 2$ (see part (c)).

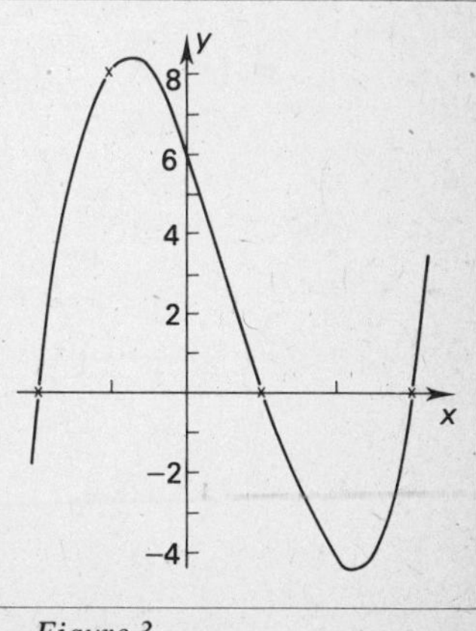

Figure 3

(c) $g'(x) = \dfrac{dg}{dx} = 3x^2 - 4x - 5 = 0 \Rightarrow x = \dfrac{4 + \sqrt{16 + 60}}{6} = 2.12$ or -0.79.

Knowing $g(x)$ is a positive cubic gives
$x = 2.12$ as a minimum point and
$x = -0.79$ as a maximum point.

Check with $g''(x) = \dfrac{d^2g}{dx^2} = 6x - 4$

$g''(2.12) = 6 \times 2.12 - 4 = 8.72 > 0 \Rightarrow x = 2.12$ is a minimum point.
$g''(-0.79) = 6 \times (-0.79) - 4 = -0.84 < 0 \Rightarrow$ a maximum point.
Points of inflexion are given by $g''(x) = 0 \Rightarrow 6x - 4 = 0 \Rightarrow x = \frac{2}{3}$.
Note that $g'(\frac{2}{3}) = 1\frac{1}{3} - 2\frac{2}{3} - 5 = -6\frac{1}{3}$ and is not zero.
Note also that in general not all points where $g''(x) = 0$ are points
of inflexion.

Example 21
Show that $(x-2)$ is a factor of $x^4 + x^3 - 7x^2 - x + 6$ and
solve the equation $x^4 + x^3 - 7x^2 - x + 6 = 0$.

To show that $x - 2$ is a factor, find the value of the expression when
$x = 2$. If this is zero, $x - 2$ is a factor.

$2^4 + 2^3 - 7 \times 2^2 - 2 + 6 = 16 + 8 - 28 - 2 + 6 = 0 \Rightarrow x - 2$ is a factor.

By division or inspection

$f(x) = x^4 + x^3 - 7x^2 - x + 6 = (x-2)(x^3 + 3x^2 - x - 3)$

Now factorize the cubic bracket; $x - 1$ is clearly a factor.

$f(x) = (x-2)(x-1)(x^2 + 4x + 3) = (x-2)(x-1)(x+1)(x+3)$

$f(x) = 0 \Rightarrow x = 2$ or $x = 1$ or $x = -1$ or $x = -3$.

This question was easy because the expression factorized. If the
cubic factor does not factorize other methods would have to be
used which might prove more difficult.

Example 22
(a) Prove that the arithmetic mean of two positive numbers
is greater than their geometric mean.

(b) Solve the inequality $\dfrac{x}{x-2} > \dfrac{1}{x+4}$

) The difficulty here is to translate the problem into algebraic
m. Let the two numbers be a and b.

Arithmetic mean, $m = \frac{1}{2}(a+b)$ Geometric mean, $g = \sqrt{(ab)}$

$m^2 - g^2 = \frac{1}{4}(a^2 + 2ab + b^2) - ab = \frac{1}{4}(a^2 + 2ab + b^2 - 4ab) = \frac{1}{4}(a-b)^2$

which is a positive quantity (the square of a real number).

$m^2 - g^2 > 0 \Rightarrow m^2 > g^2 \Rightarrow m > g$ (since m and g are positive).

The expression for g is difficult to cope with algebraically and that is why we work with g^2.

(b) Inequalities are treated just like equalities (equations) but with one important difference: when multiplying both sides by a negative number the inequality sign is reversed. Since x, $x-2$ and $x+4$ can be negative, we consider different sets of x values.

$x > 2 \Rightarrow x(x+4) > x-2 \Rightarrow x^2 + 3x + 2 > 0 \Rightarrow (x+2)(x+1) > 0$

The inequality is satisfied when $x > 2$.

$$0 < x < 2 \Rightarrow \frac{x}{x-2} < 0 \text{ and } \frac{1}{x+4} > 0 \Rightarrow \text{no solution}$$

$-4 < x < 0 \Rightarrow x-2 < 0$ and $x(x+4) < x-2$

(inequality sign changes)

$$\Rightarrow x^2 + 4x < x - 2 \Rightarrow x^2 + 3x + 2 < 0 \Rightarrow (x+1)(x+2) < 0$$

This is satisfied when $-2 < x < -1$.

$$x < -4 \Rightarrow \frac{x}{x-2} > 0, \frac{1}{x+4} < 0 \text{ and the inequality is satisfied.}$$

Complete solution: $x < -4$, $-2 < x < -1$, $x > 2$.

The above method of solution is not easy to follow and certainly difficult to apply. The following method is much better.

$$\frac{x}{x-2} > \frac{1}{x+4} \Rightarrow \frac{x}{x-2} - \frac{1}{x+4} > 0 \Rightarrow \frac{x(x+4)-(x-2)}{(x-2)(x+4)} > 0$$

$$\Rightarrow E = \frac{x^2 + 3x + 2}{(x-2)(x+4)} = \frac{(x+1)(x+2)}{(x-2)(x+4)} > 0$$

Boundaries or critical x values		-4		-2		-1		2	
Values for E on each side of these critical x values are		$+$ve		$-$		$+$		$-$		$+$

Solution $x < -4$; $-2 < x < -1$; $x > 2$

Example 23

Find the centre of symmetry of the curve
$y = x^3 - 3x^2 - 4x + 12$.

The graph is a positive cubic.
Will it factorize?

$$y = x^2(x-3) - 4(x-3)$$
$$= (x^2 - 4)(x-3)$$
$$= (x+2)(x-2)(x-3)$$

The graph cuts the x-axis at -2, 2 and 3.

The centre of symmetry will be the central point of inflexion. To find this

$$\frac{dy}{dx} = 3x^2 - 6x - 4$$

$$\frac{d^2y}{dx^2} = 6x - 6 = 0 \Rightarrow x = 1$$

$$x = 1 \Rightarrow y = 3(-1)(-2) = 6$$

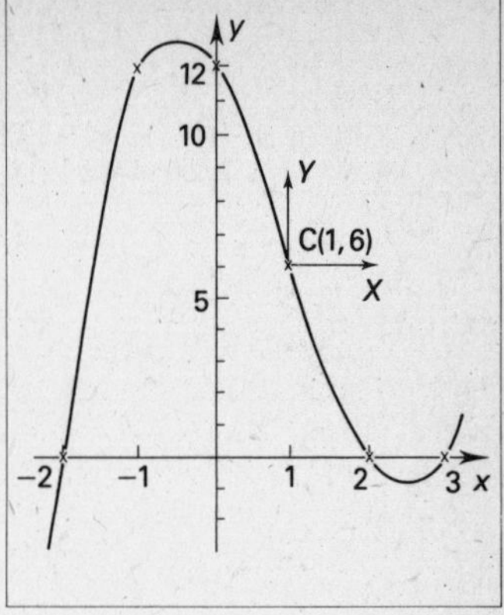

Figure 4

To confirm the centre of symmetry $C(1,6)$ refer the curve to coordinate axes X, Y centred at C.
This means $x = X+1$ and $y = Y+6$.
Substitute for x and y in the original equation.

$$Y+6 = (X+1)^3 - 3(X+1)^2 - 4(X+1) + 12$$
$$= X^3 + 3X^2 + 3X + 1 - 3X^2 - 6X - 3 - 4X - 4 + 12$$
$$Y = X^3 - 7X$$

Referred to X, Y axes this equation represents an **odd** function. This means it has rotational symmetry about $X = 0$, $Y = 0$, i.e., the point C. C is then the centre of symmetry of the original curve.

Example 24
Solve the inequalities
(a) $2(x-1) > x+1$ (b) $2|x-1| > |x+1|$
(c) $\dfrac{1}{x-1} > \dfrac{1}{x+1}$

(a) $2(x-1) > x+1 \Rightarrow 2x-2 > x+1 \Rightarrow x > 3$

(b) $|x-1|$ means the size of $x-1$ irrespective of sign (positive or negative).
When $x > 1$, $|x-1| = x-1$ and when $x < 1$, $|x-1| = 1-x$.
For $x > 1$, $2|x-1| > |x+1| \Rightarrow 2(x-1) > x+1 \Rightarrow x > 3$
For $-1 < x < 1$, $2|x-1| > |x+1| \Rightarrow 2(1-x) > x+1 \Rightarrow 1 > 3x$
$$\Rightarrow x < \tfrac{1}{3}$$

For $\quad x < -1, \quad 2|x-1| > |x+1| \Rightarrow 2(1-x) > (-x-1)$
$$\Rightarrow 3 > x$$

Complete solution $\quad x > 3$ or $x < \frac{1}{3}$ (see Figure 5(a)).

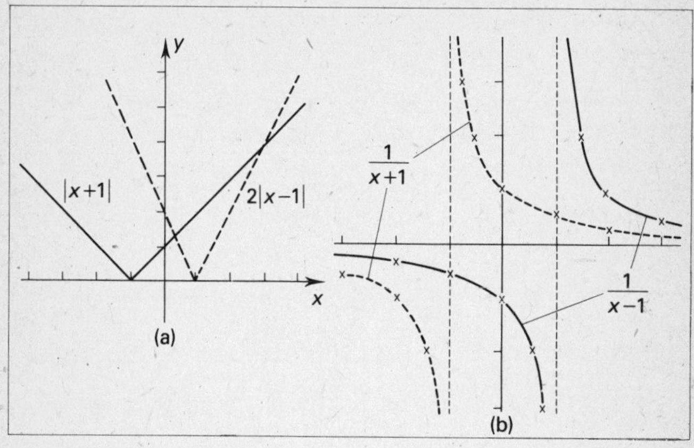

Figure 5

(c) For $x > 1$ and $x < -1$ the inequality sign stays the same:

$$\frac{1}{x-1} > \frac{1}{x+1} \Rightarrow x+1 > x-1 \Rightarrow 2 > 0 \text{ which is true for all } x.$$

For $-1 < x < 1, \quad \dfrac{1}{x-1} > \dfrac{1}{x+1} \Rightarrow x+1 < x-1 \Rightarrow 2 < 0$

which is false.

No values within this set satisfy the inequality.
Complete solution $\quad x > 1$ or $x < -1$ (see Figure 5(b)).

Alternatively $\quad \dfrac{1}{x-1} - \dfrac{1}{x+1} > 0 \Rightarrow \dfrac{2}{(x-1)(x+1)} > 0$

$$\Rightarrow x > 1 \text{ or } x < -1.$$

These inequalities can be solved algebraically, but confirmation of the results and greater understanding can be obtained by drawing graphs. Figure 5(a) shows the graphs of $y = |x+1|$ and $y = 2|x-1|$ (dotted) and the dotted graph lies above the full lined graph when $x > 3$ or $x < \frac{1}{3}$.

The result for part (c) is shown by the graphs of $y = (x-1)^{-1}$ and $y = (x+1)^{-1}$ in Figure 5(b).

Chapter 2

Trigonometry

Multiple choice questions (Type A) Select the correct answer

Example 1

$\sin (A+B) - \sin (A-B) =$

A $2 \sin A \cos B$ **B** $2 \sin A \sin B$ **C** $2 \cos A \sin B$
D $2 \cos A \cos B$ **E** $-2 \cos A \sin B$

Method 1 $\sin (A+B) = \sin A \cos B + \cos A \sin B$
$\sin (A-B) = \sin A \cos B - \cos A \sin B$

Subtracting these two equations gives $2 \cos A \sin B$.   ANSWER C

Method 2   Using $\sin P - \sin Q = 2 \cos \frac{1}{2}(P+Q) \sin \frac{1}{2}(P-Q)$
$P+Q = (A+B)+(A-B) = 2A$ and $P-Q = (A+B)-(A-B) = 2B$
So $\sin (A+B) - \sin (A-B) = 2 \cos A \sin B$.   ANSWER C

In this question there is little to choose between the two methods. However, in some trigonometrical questions, using the right formula at the right time can save much time and working. It is advisable to learn these formulae thoroughly.

Example 2

$\cos 120° =$

A $\dfrac{\sqrt{3}}{2}$ **B** $\frac{1}{2}$ **C** $-\dfrac{\sqrt{3}}{2}$ **D** $-\frac{1}{2}$ **E** $\dfrac{1}{\sqrt{2}}$

$\cos 120° = -\cos (180° - 120°) = -\cos 60° = -\frac{1}{2}$.   ANSWER D

Remember to take the angle away from 180° and that the cosine is negative in the second quadrant.

Example 3

If x is acute, $\tan x = \dfrac{a}{b} \Rightarrow \cos x =$

A $\dfrac{b}{a}$ **B** $\dfrac{a}{\sqrt{a^2+b^2}}$ **C** $\dfrac{b}{\sqrt{a^2+b^2}}$ **D** $\dfrac{a}{\sqrt{a^2-b^2}}$ **E** $\dfrac{b}{\sqrt{a^2-b^2}}$

Draw a triangle (right-angled) with x as an acute angle, a as the opposite side and b as the adjacent side. Then the hypotenuse is $\sqrt{a^2+b^2}$ and $\cos x = b/\sqrt{a^2+b^2}$. 　　　　　ANSWER C

Example 4

If $f(x) = 3\cos x + 4\sin x$ then the maximum value of $f(x)$ is

A 3　**B** 4　**C** 5　**D** 6　**E** 7

Method 1　$f'(x) = \dfrac{df}{dx} = -3\sin x + 4\cos x = 0$ for maximum.

$\Rightarrow 3\sin x = 4\cos x$

$\Rightarrow \tan x = \frac{4}{3}$ 　　　　　(since $\sin x \div \cos x = \tan x$)

$\Rightarrow \sin x = \frac{4}{5}$ and $\cos x = \frac{3}{5}$. There is no need to find x.

$f(\max) = \frac{9}{5} + \frac{16}{5} = \frac{25}{5} = 5$. 　　　　　ANSWER C

Method 2　$f(x) = 5(\frac{3}{5}\cos x + \frac{4}{5}\sin x)$

Put $\sin y = \frac{3}{5}$ and $\cos y = \frac{4}{5}$,

$$f(x) = 5(\sin y \cos x + \cos y \sin x)$$

$$= 5\sin(y+x)$$

This will give a maximum value of 5 since the maximum value of $\sin(y+x)$ is 1. 　　　　　ANSWER C

Method 1 is the obvious way to find the maximum value of a function but Method 2 has an important application when solving equations (see Example 28).

Example 5

$\dfrac{d}{dx}(\ln \cos x) =$

A $\dfrac{1}{\cos x}$ 　　**B** $\dfrac{\sin x}{\cos x}$ 　　**C** $-\dfrac{1}{\sin x}$ 　　**D** $-\tan x$ 　　**E** $\sec x$

This is a composite function (or function of a function). The derivative of $\ln y$ is $\dfrac{1}{y}$ and in this example the logarithm is differentiated first to give $\dfrac{1}{\cos x}$ which is then multiplied by the derivative of $\cos x$, namely $-\sin x$.

The result is $\dfrac{-\sin x}{\cos x} = -\tan x.$ ANSWER D

Some students forget the way to differentiate a composite function and produce answer A (answer E is the same as answer A) or confuse the two processes (differentiating the log and the cos) to produce answer C.

Example 6
$\tan(60° + 45°) =$

A $\dfrac{\sqrt{3}+1}{1-\sqrt{3}}$ **B** $\dfrac{1+\sqrt{3}}{\sqrt{3}-1}$ **C** $\dfrac{\sqrt{3}-1}{\sqrt{3}+1}$ **D** $\dfrac{1-\sqrt{3}}{1+\sqrt{3}}$

E none of these

$$\tan(60° + 45°) = \frac{\tan 60° + \tan 45°}{1 - \tan 60° \tan 45°} = \frac{\sqrt{3}+1}{1 - \sqrt{3} \times 1}. \qquad \text{ANSWER A}$$

It is worthwhile learning the ratios associated with the 30°, 60°, 90° and 45°, 45°, 90° triangles. Examiners often ask questions on these triangles.

$$\sin 30° = \cos 60° = \tfrac{1}{2}; \quad \sin 60° = \cos 30° = \frac{\sqrt{3}}{2}; \quad \tan 30° = \frac{1}{\sqrt{3}};$$

$$\tan 60° = \sqrt{3}$$

Multiple choice questions (Type B) Answer according to the table

A	**B**	**C**	**D**	**E**
1, 2, 3	1, 3	2, 3	2	3
correct	only	only	only	only

Example 7
Given that $\tan A = t$

1 $\sin 2A = \dfrac{2t}{1+t^2}$ **2** $\tan 2A = \dfrac{1+t^2}{1-t^2}$ **3** $\cos 2A = \dfrac{1-t^2}{1+t^2}$

$$\sin 2A = 2\sin A \cos A = \frac{2\sin A \cos A}{\cos^2 A + \sin^2 A}$$

Dividing top and bottom by $\cos^2 A$

$$= \frac{2\tan A}{1+\tan^2 A} \qquad \text{so statement 1 is correct}$$

$$\cos 2A = \cos^2 A - \sin^2 A = \frac{\cos^2 A - \sin^2 A}{\cos^2 A + \sin^2 A}$$

$$= \frac{1-\tan^2 A}{1+\tan^2 A} \qquad \text{so statement 3 is correct}$$

$$\tan 2A = \frac{\sin 2A}{\cos 2A} = \frac{2t}{1-t^2} \qquad \text{so statement 2 is wrong}$$

ANSWER B

Example 8

$\cos 2A =$

1 $\cos^2 A + \sin^2 A$ **2** $2\cos^2 A - 1$ **3** $1 - 2\sin^2 A$

$\cos(A+B) = \cos A \cos B - \sin A \sin B$ and putting $B = A$ gives

$$\begin{aligned}
\cos 2A &= \cos^2 A - \sin^2 A \\
&= 1 - \sin^2 A - \sin^2 A \\
&= 1 - 2\sin^2 A \\
&= 1 - 2(1 - \cos^2 A) \\
&= 2\cos^2 A - 1
\end{aligned}$$

so statement 1 is wrong
since $\cos^2 A = 1 - \sin^2 A$
so statement 3 is correct
since $\sin^2 A = 1 - \cos^2 A$
so statement 2 is correct

ANSWER C

Example 9

If $180° < A < 270°$ then

1 $\sin A$ is positive   **2** $\cos A$ is positive   **3** $\tan A$ is positive

In the third quadrant $\tan A$ is the only ratio which is positive.

ANSWER E

Most students devise a way for remembering which ratios are positive in the four quadrants.

For $0° < A < 90°$ in the first quadrant **All** ratios are **Positive**
For $90° < A < 180°$ in the second quadrant the **Sine** is **Positive**
For $180° < A < 270°$ in the third quadrant the **Tangent** is **Positive**
For $270° < A < 360°$ in the fourth quadrant the **Cosine** is **Positive**

> **Example 10**
>
> A curve is given parametrically by $x = \cos t$, $y = \cos 2t$.
>
> **1** The curve is symmetrical about the y-axis.
> **2** The curve is a parabola.
> **3** Its Cartesian equation is $y = 2x^2 - 1$.

Consider statement 3 first: $y = \cos 2t = 2\cos^2 t - 1 = 2x^2 - 1$ which is symmetrical about the y-axis, but the values for x and y lie between -1 and $+1$ so the curve is that part of the parabola lying within the square $-1 \leqslant x \leqslant +1$ and $-1 \leqslant y \leqslant +1$. Statement 2 is therefore only partly true. $\hfill$ ANSWER B

> **Example 11**
>
> $I = \int 2 \sin x \cos x \, dx =$
>
> **1** $\sin^2 x + c$ $\quad$ **2** $-\cos^2 x + k$ $\quad$ **3** $-\tfrac{1}{2} \cos 2x + K$

Substituting $y = \sin x$ gives $dy = \cos x \, dx$ so

$\quad I = \int 2y \, dy = y^2 + c = \sin^2 x + c \hfill$ statement 1 is correct

Substituting $y = \cos x$ gives $dy = -\sin x \, dx$ so

$\quad I = \int -2y \, dy = -y^2 + k = -\cos^2 x + k \hfill$ statement 2 is correct

$\quad I = \int \sin 2x \, dx = -\tfrac{1}{2} \cos 2x + K \hfill$ statement 3 is correct

All three answers are correct. $\hfill$ ANSWER A

At first these results puzzle the student but the three answers only differ from each other by a constant amount.

$$\sin^2 x = 1 - \cos^2 x \quad \text{and} \quad -\tfrac{1}{2}\cos 2x = -\tfrac{1}{2}(2\cos^2 x - 1) = \tfrac{1}{2} - \cos^2 x$$

When the limits of integration are inserted all three give the same result.

> **Example 12**
>
> In comparison with $y = \sin x$ the graph of
>
> **1** $y = 2\sin x$ has double the frequency.
> **2** $y = \sin 2x$ has double the amplitude.
> **3** $y = 2\sin 2x$ has double the amplitude and double the frequency.

In statement 1, each value of $\sin x$ is doubled so the graph will appear twice as high, i.e., it will take values between -2 and $+2$ while $y = \sin x$ lies between -1 and $+1$. Its amplitude is doubled. $y = \sin x$ completes a full cycle while x changes by 2π radians ($360°$) whereas $y = \sin 2x$ will complete a full cycle while x changes by π radians ($180°$). So $\sin 2x$ completes two cycles while $\sin x$ is completing only one and so has double the frequency. Statements 1 and 2 are both wrong being the wrong way round, and statement 3 is correct.   ANSWER E

Example 13

The factors of $\sin 3A + \sin 5A$ are

1 $\sin A$ and $\sin 4A$ **2** $\sin A$ and $\cos 4A$ **3** $\cos A$ and $\sin 4A$

The required factor formula is

$$\sin P + \sin Q = 2 \sin \tfrac{1}{2}(P+Q) \cos \tfrac{1}{2}(P-Q)$$

which gives

$$\sin 3A + \sin 5A = 2 \sin \tfrac{1}{2}(3A+5A) \cos \tfrac{1}{2}(3A-5A)$$
$$= 2 \sin 4A \cos(-A) = 2 \sin 4A \cos A$$

So the factors are in statement 3.   ANSWER E

Example 14

The graph of $y = \cos^2 x$

1 is periodic   **2** has amplitude 1   **3** is symmetrical

$\cos 2x = 2\cos^2 x - 1 \Rightarrow \cos^2 x = \tfrac{1}{2}(1 + \cos 2x)$ which has period π radians ($180°$) and varies between 0 and 1 and so has amplitude $\tfrac{1}{2}$. Being a transformation of $\cos x$ it has vertical lines of symmetry and rotational symmetry. Therefore statements 1 and 3 are correct.   ANSWER B

Multiple choice questions (Type C) For each question two statements are given. Answer

A if 1 implies 2 and 2 implies 1
B if 2 implies 1 but 1 does not imply 2
C if 1 implies 2 but 2 does not imply 1
D if 1 denies 2 and 2 denies 1
E if none of these relations hold

Students find the logic for answering this type of question difficult to grasp. The idea is to emphasize the concept of necessary and sufficient conditions. In this chapter some of the questions are easy but choosing the correct implication sign shows whether the logic of each statement is reversible.

Example 15

$$1 \quad \frac{d}{dx}(a\cos x + b\sin x) = 0 \qquad 2 \quad \tan x = \frac{b}{a}$$

If statement 1 is true then $-a\sin x + b\cos x = 0 \Rightarrow a\sin x = b\cos x$

$$\Rightarrow \tan x = \frac{b}{a}$$

So statement 1 implies statement 2 and 2 implies 1.   ANSWER A

Example 16

If A and B are angles between $0°$ and $90°$

$$1 \quad A > B \qquad 2 \quad \cos A > \cos B$$

For $0° < x < 90°$, $\cos x$ is a decreasing function so within this interval if A is larger than B then $\cos A$ will be smaller than $\cos B$. Statement 1 denies statement 2 and 2 denies 1.   ANSWER D

Example 17

$$1 \quad y = \sin x \qquad 2 \quad \frac{dy}{dx} = \cos x$$

If $y = \sin x$ then $\frac{dy}{dx} = \cos x$ but $y = 6 + \sin x \Rightarrow \frac{dy}{dx} = \cos x$. So statement 1 implies statement 2 but 2 does not imply 1, because to find y from $\frac{dy}{dx}$ it is necessary to integrate, which introduces a constant.   ANSWER C

Multiple choice questions (Type D) Each question consists of a problem followed by four pieces of information. Decide whether the problem can be solved with one of the four pieces of information omitted and answer

A if 1 could be omitted   **D** if 4 could be omitted
B if 2 could be omitted   **E** if none can be omitted
C if 3 could be omitted

Example 18

Find $\displaystyle\int_0^{\pi/2} (P\cos Qx + R\sin Sx)\,dx$

1 $P = 12$ **2** $Q = 32$ **3** $R = 5$ **4** $S = 87$

$$\int_0^{\pi/2} (P\cos Qx + R\sin Sx)\,dx = \left[\frac{P}{Q}\sin Qx - \frac{R}{S}\cos Sx\right]_0^{\pi/2}$$

$$= \left[\frac{P}{Q}\sin\tfrac{1}{2}Q\pi - \frac{R}{S}\cos\tfrac{1}{2}S\pi + \frac{R}{S}\right]$$

$\sin\tfrac{1}{2}Q\pi = \sin 16\pi = 0$;
1 is not needed.   ANSWER A

Short questions

Example 19
Expand $\sin x \cos x$ as a series in ascending powers of x as
far as the term in x^3. Compare this with the series for $\sin 2x$.

$\sin x = x - \tfrac{1}{6}x^3 + \dots$ and $\cos x = 1 - \tfrac{1}{2}x^2 + \dots$

$\sin x \cos x = (x - \tfrac{1}{6}x^3)(1 - \tfrac{1}{2}x^2) = x - \tfrac{1}{2}x^3 - \tfrac{1}{6}x^3 + \dots = x - \tfrac{2}{3}x^3$

$\sin 2x = 2x - \dfrac{(2x)^3}{3!} = 2x - \dfrac{4x^3}{3}$

This is double the series of $\sin x \cos x$, which is just what we expect
as $\sin 2x = 2\sin x \cos x$.

Example 20
Given that $\tan x = \tfrac{5}{12}$ and $0° < x < 90°$ find $\tan 2x$, $\sin 2x$
and $\cos 2x$.

$\tan x = \tfrac{5}{12} \Rightarrow \sin x = \tfrac{5}{13}$ and $\cos x = \tfrac{12}{13}$. These values can be found
easily by drawing a right-angled triangle, putting 5 on the side
opposite angle x and 12 adjacent and giving a hypotenuse of 13 (by

Pythagoras). The sine and cosine ratios can then be calculated. This process is used quite often, especially with $3, 4, 5$ and $5, 12, 13$ triangles.

$$\tan 2x = \frac{2\tan x}{1-\tan x} = \frac{\frac{5}{6}}{1-\frac{25}{144}} = \frac{\frac{5}{6}}{\frac{119}{144}} = \frac{120}{119}$$

$$\sin 2x = 2\sin x \cos x = 2 \times \frac{5}{13} \times \frac{12}{13} = \frac{120}{169}$$

$$\cos 2x = \cos x - \sin x = \frac{144}{169} - \frac{25}{169} = \frac{119}{169}$$

Incidentally we have found another Pythagorean triple 119, 120, 169 making a right-angled triangle with integer (whole number) sides, whose angles are very close to $45°, 45°, 90°$.

Example 21

Given that $\cos 2x = \frac{119}{169}$ and $0° < x < 90°$ find the value of $\cos x$ without using tables or a calculator.

$$\cos 2x = 2\cos^2 x - 1 = \frac{119}{169} \Rightarrow 2\cos^2 x = 1 + \frac{119}{169} = \frac{288}{169}$$

$$\Rightarrow \quad \cos^2 x = \frac{144}{169} \Rightarrow \cos x = \frac{12}{13}$$

This can be seen from Example 20 to be correct.

Example 22

Without using tables find the values of $\sin 15°$, $\cos 15°$ and $\tan 15°$, leaving your answers in surd form.

$$\sin 15° = \sin(45° - 30°) = \sin 45° \cos 30° - \cos 45° \sin 30°$$

$$= \frac{1}{\sqrt{2}} \times \frac{\sqrt{3}}{2} - \frac{1}{\sqrt{2}} \times \frac{1}{2} = \frac{\sqrt{3}-1}{2\sqrt{2}}$$

$$\cos 15° = \cos(45° - 30°) = \cos 45° \cos 30° + \sin 45° \sin 30°$$

$$= \frac{1}{\sqrt{2}} \times \frac{\sqrt{3}}{2} + \frac{1}{\sqrt{2}} \times \frac{1}{2} = \frac{\sqrt{3}+1}{2\sqrt{2}}$$

$$\tan 15° = \frac{\sin 15°}{\cos 15°} = \frac{\sqrt{3}-1}{\sqrt{3}+1}$$

It is worthwhile building in checks to your working to make sure you do not make silly mistakes. As an example, $(\sqrt{3}-1)$, $(\sqrt{3}+1)$ and $2\sqrt{2}$ should form the sides of a right-angled triangle, and can be checked by Pythagoras.

Example 23
Find the mean value of the function $\sin x$ over the interval $0 < x < \pi/2$.

The mean value of a function is given by the integral over that interval divided by the length of the interval.

$$\text{Mean value} = \frac{\int \sin x \, dx}{\frac{1}{2}\pi} = \frac{2}{\pi}[-\cos x]_0^{\pi/2}$$

$$= \frac{2}{\pi}(-\cos \pi/2 + \cos 0) = \frac{2}{\pi}$$

It is worth remembering that the area under the graph of $\sin x$ or $\cos x$ from 0 to $\pi/2$ is 1.

Example 24
Find the mean value of the function $y = \cos^2 x$ for $0 \leqslant x \leqslant \pi$.

$$\text{Mean value} = \frac{1}{\pi} \int \cos^2 x \, dx$$

$$= \frac{1}{\pi} \int_0^\pi \tfrac{1}{2}(1 + \cos 2x) \, dx$$

$$= \frac{1}{2\pi} [x + \tfrac{1}{2}\sin 2x]_0^\pi$$

$$= \frac{1}{2\pi} \times \pi = \tfrac{1}{2}$$

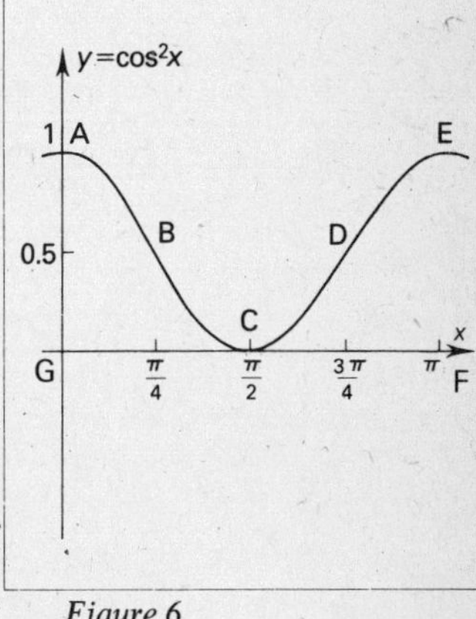

Figure 6

There is no need to calculate the integral of $\cos x$. This can be done using the symmetry of the graph in Figure 6. For every value of the function from 0° to 45° (A to B) there is a value from 45° to 90° (B to C) which complements it so that the sum of the two values is 1 and their average is $\tfrac{1}{2}$. This also applies to the values between CD and DE.

Example 25
A watering trough is 1 m long and its cross-section is in the form of the graph of $y = \cos^2 x$ for $0 \leqslant x \leqslant \pi$. If the trough is 60 cm wide and 30 cm deep, find its capacity.

We need to find the area of cross-section which will be similar in shape to the graph in Figure 6. The graph $ABCDE$ divides the rectangle $AEFG$ in half and the rectangle has area $AG \times GF = 1 \times \pi$. The cross-sectional area of the trough will be $\frac{1}{2} \times 30 \times 60 \,\text{cm}^2$.
Volume of trough $= \frac{1}{2} \times 30 \times 60 \times 100 \,\text{cm}^3 = 90\,000 \,\text{cc} = 90 \,\text{litres}$.

Example 26

What are the periods of

   (a) $\cos x$ (b) $\cos^2 x$ (c) $\cos^3 x$

(a) The period is the time (or length) of one oscillation, or the difference in x values between peak values (or troughs). For $\cos x$ this is 2π radians.

(b) From Figure 6 the graph of $\cos^2 x$ completes one cycle in π^c so its period is π^c.

(c) $\cos 3x = 4\cos^3 x - 3\cos x \Rightarrow \cos^3 x = \frac{3}{4}\cos x + \frac{1}{4}\cos 3x$.

$\text{Cos } x$ has period 2π, $\cos 3x$ has period $\dfrac{2\pi}{3}$ and repeats its values

for the third time every $2\pi^c$. $\text{Cos}^3 x$ will not repeat values until $\cos x$ does, i.e. after $2\pi^c$. Its period is $2\pi^c$.

Long questions

Example 27

(a) Show that $\sin 3A = 3\sin A - 4\sin^3 A$

(b) Hence, or otherwise, solve $\sin 3A = \sin^2 A$

(a) $\sin 3A = \sin(2A + A) = \sin 2A \cos A + \cos 2A \sin A$

$$= (2\sin A \cos A)\cos A + (1 - 2\sin^2 A)\sin A$$
$$= 2\sin A\,(1 - \sin^2 A) + \sin A - 2\sin^3 A$$
$$= 2\sin A - 2\sin^3 A + \sin A - 2\sin^3 A$$
$$= 3\sin A - 4\sin^3 A$$

(b) $\sin 3A = \sin^2 A \Rightarrow 3\sin A - 4\sin^3 A = \sin^2 A$

$$\Rightarrow 4\sin^3 A + \sin^2 A - 3\sin A = 0$$
$$\Rightarrow \sin A(4\sin^2 A + \sin A - 3) = 0$$
$$\Rightarrow \sin A(\sin A + 1)(4\sin A - 3) = 0$$

$\Rightarrow \sin A = 0$ or $\sin A = -1$ or $\sin A = \frac{3}{4}$

$\Rightarrow A = 0°, 180°, 360°$ or $A = 270°$ or $A = 48.6°, 131.4°$

Example 28

Solve the equation $3 \sin A + 4 \cos A = 2$ for $0° \leqslant A \leqslant 360°$

Method 1 $\quad \frac{3}{5} \sin A + \frac{4}{5} \cos A = \frac{2}{5}$

Write $\cos B = \frac{3}{5}$ and $\sin B = \frac{4}{5} \Rightarrow B = 53.1°$
So $\quad \cos B \sin A + \sin B \cos A = 0.4$
$\quad \quad \sin (A + B) = 0.4 \Rightarrow A + B = 23.6°$ or $156.4° + 360k°$

$$B = 53.1 \Rightarrow A = 23.6° - 53.1° + 360k° \quad \text{or} \quad 156.4° - 53.1° + 360k°$$
$$= 360° - 29.5° + 360k° \quad \text{or} \quad 103.3° \quad \quad + 360k°$$
$$= 330.5° \quad \quad \quad + 360k° \quad \text{or} \quad 103.3° \quad \quad \quad + 360k°$$

Method 2 If $t = \tan \frac{1}{2}A$ then $\sin A = \dfrac{2t}{1+t^2}$ and $\cos A = \dfrac{1-t^2}{1+t^2}$

Substituting these in the equation gives $6t + 4(1 - t^2) = 2(1 + t^2)$
$\Rightarrow 6t + 4 - 4t^2 = 2 + 2t^2$
$\Rightarrow 6t^2 - 6t - 2 = 0 \Rightarrow 3t^2 - 3t - 1 = 0$

$$\Rightarrow t = \frac{3 \pm \sqrt{9+12}}{6} = \frac{3 \pm 4.5826}{6} = 1.2638 \text{ or } -0.2638$$

$\tan \frac{1}{2}A = 1.2638 \quad \Rightarrow \frac{1}{2}A = 51.65° \quad \Rightarrow A = 103.3°$
$\tan \frac{1}{2}A = -0.2638 \Rightarrow \frac{1}{2}A = -14.78° \Rightarrow A = -29.5°$ or $330.5°$
General solution $A = 103.3° + 360k°$ or $330.5° + 360k°$.

Example 29

Show that for all A, $\cos A + \cos (A + 120°) + \cos (A + 240°) = 0$.

Method 1 $\quad \cos (A + 120°) = \cos A \cos 120° - \sin A \sin 120°$

$$= -\tfrac{1}{2} \cos A - \frac{\sqrt{3}}{2} \sin A$$

$\cos (A + 240°) = \cos A \cos 240° - \sin A \sin 240°$

$$= -\tfrac{1}{2} \cos A + \frac{\sqrt{3}}{2} \sin A$$

$\cos A + \cos (A + 120°) + \cos (A + 240°) = 0$

Method 2 Use $\cos P + \cos Q = 2 \cos \frac{1}{2}(P + Q) \cos \frac{1}{2}(P - Q)$ to give

$\cos A + \cos (A + 240°) = 2 \cos \frac{1}{2}(2A + 240°) \cos \frac{1}{2}(-240°)$
$\quad \quad \quad \quad \quad \quad = -\cos (A + 120°) \quad$ since $\cos (-120°) = -\frac{1}{2}$

Hence $\quad \cos A + \cos (A + 240°) + \cos (A + 120°) = 0$

Method 3 (See Figure 7 for vector diagram.) *PQR* is an equilateral triangle with *PQ* making an angle of *A* with the positive *x*-axis.

Vector $\mathbf{PQ} = \begin{pmatrix} \cos A \\ \sin A \end{pmatrix}$

or $\cos A\mathbf{i} + \sin A\mathbf{j}$

QR makes an angle of $(A+120°)$ with the positive *x*-axis, so

$$\mathbf{QR} = \begin{pmatrix} \cos (A+120°) \\ \sin (A+120°) \end{pmatrix}$$

and

$$\mathbf{RP} = \begin{pmatrix} \cos (A+240°) \\ \sin (A+240°) \end{pmatrix}$$

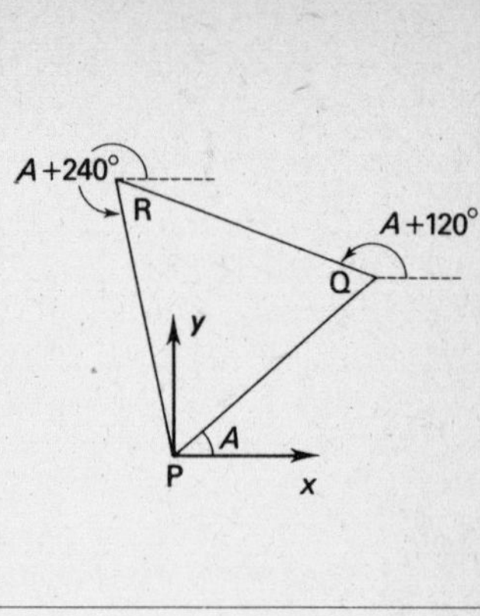

Figure 7

The vectors representing the sides of the triangle add to zero, i.e.,

$$\mathbf{PQ} + \mathbf{QR} + \mathbf{RP} = 0$$

$$\begin{pmatrix} \cos A \\ \sin A \end{pmatrix} + \begin{pmatrix} \cos (A+120°) \\ \sin (A+120°) \end{pmatrix} + \begin{pmatrix} \cos (A+240°) \\ \sin (A+240°) \end{pmatrix} = \begin{pmatrix} 0 \\ 0 \end{pmatrix}$$

This gives the required result and a similar result for sines.

In this example, the student will probably find Method 1 the easiest but Method 2 illustrates how useful the factor formulae can be and Method 3 shows how vectors can be used. Use of the methods of matrices and vectors is the best way of deriving the addition formulae used in Method 1.

Example 30

On the same diagram sketch the curves whose polar equations are

$$C \quad r = 1 + \cos \theta; \qquad C' \quad r = 2 + \cos \theta$$

Find the area enclosed between the two curves.

Both curves are symmetrical about the *x*-axis (see Figure 8(a)). The area enclosed by a curve is given by $\int \frac{1}{2} r^2 \theta \, d\theta$.

For C

$$A = 2 \int_0^{\pi} \tfrac{1}{2}(1+\cos\theta)^2 \, d\theta = \int_0^{\pi} (1+2\cos\theta+\cos^2\theta) \, d\theta$$

$$= \int_0^{\pi} (1+2\cos\theta+\tfrac{1}{2}(1+\cos 2\theta)) \, d\theta$$

$$= [1\tfrac{1}{2}\theta+2\sin\theta+\tfrac{1}{4}\sin 2\theta]_0^{\pi} = 1\tfrac{1}{2}\pi$$

since $\sin\pi = \sin 2\pi = \sin 0 = 0$.

For C'

$$A = 2 \int_0^{\pi} \tfrac{1}{2}(2+\cos\theta)^2 \, d\theta = \int_0^{\pi} (4+4\cos\theta+\cos^2\theta) \, d\theta = 4\tfrac{1}{2}\pi$$

Area between the curves $= 4\tfrac{1}{2}\pi - 1\tfrac{1}{2}\pi = 3\pi.$

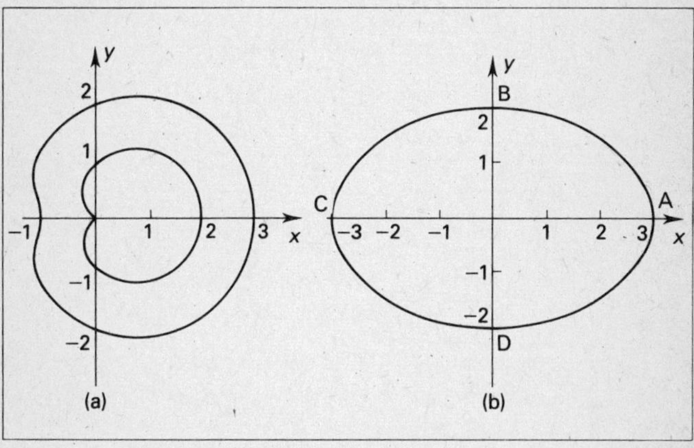

Figure 8

As a check on these two integrals, C is a little larger than the circle centre $(1,0)$, radius 1 and area π, and C' is a little larger than the circle centre $(1,0)$, radius 2 and area 4π.

Chapter 3

Differentiation

Differentiation is the name given to the process by which the rate of change of one variable with respect to another can be found.

Clearly it is important that the basic concepts of differentiation should be thoroughly understood before attempting to apply the ideas to particular problems.

There are a number of fundamental methods and techniques for evaluating the differentials of more complex functions. Those required by most syllabuses include the product and quotient rules, the function of a function method, implicit and parametric differentiation and logarithmic differentiation.

These basic techniques can be used on algebraic, trigonometrical, exponential, logarithmic, hyperbolic and inverse functions. Questions may be set which simply test these processes or their application to various problems such as turning points of a curve, velocity and acceleration, small changes, etc.

Multiple choice questions (Type A) Select the correct answer

Example 1

If $y = x/\sin x$ then $\dfrac{dy}{dx} =$

A $\dfrac{\sin x - x \cos x}{x^2}$ **B** $\dfrac{\sin x + x \cos x}{\sin^2 x}$ **C** $\dfrac{\sin x - x \cos x}{\sin^2 x}$

D $\dfrac{x \cos x - \sin x}{\sin^2 x}$ **E** $\dfrac{x \cos x - \sin x}{x^2}$

Since the given function is a quotient in the form $y = \dfrac{u}{v}$ the solution can be found by using the quotient rule.

$$\frac{d}{dx}\left(\frac{u}{v}\right) = \frac{v\left(\dfrac{du}{dx}\right) - u\left(\dfrac{dv}{dx}\right)}{v^2}$$

Thus $u = x \Rightarrow \dfrac{du}{dx} = 1$ and $v = \sin x \Rightarrow \dfrac{dv}{dx} = \cos x$

If $y = \dfrac{x}{\sin x}$ then $\dfrac{dy}{dx} = \dfrac{1 \cdot \sin x - x \cdot \cos x}{\sin^2 x}$

$$= \dfrac{\sin x - x \cos x}{\sin^2 x} \qquad \text{ANSWER C}$$

Example 2

If $y = \ln\left(\cos\dfrac{x}{3}\right)$ then $\dfrac{dy}{dx} =$

A $\;-\tfrac{1}{3}\tan\left(\dfrac{x}{3}\right)$ **B** $\;-\tan\left(\dfrac{x}{3}\right)$ **C** $\;-\dfrac{1}{\cos\left(\dfrac{x}{3}\right)}$

D $\;-\dfrac{1}{3\cos\left(\dfrac{x}{3}\right)}$ **E** $\;-\dfrac{3}{\cos\left(\dfrac{x}{3}\right)}$

If we let $u = \cos\left(\dfrac{x}{3}\right)$ then $y = \ln u$.

Hence y is a function of u, which is itself a function of x and thus the function of a function result can be used.

If $u = \cos\left(\dfrac{x}{3}\right)$ then $\dfrac{du}{dx} = -\tfrac{1}{3}\sin\left(\dfrac{x}{3}\right)$

If $y = \ln u$ then $\dfrac{dy}{du} = \dfrac{1}{u} = \dfrac{1}{\cos\left(\dfrac{x}{3}\right)}$

As $\dfrac{dy}{dx} = \dfrac{dy}{du} \times \dfrac{du}{dx}$ we have $\dfrac{dy}{dx} = -\tfrac{1}{3}\sin\left(\dfrac{x}{3}\right) \times \dfrac{1}{\cos\left(\dfrac{x}{3}\right)}$

$$= -\tfrac{1}{3}\tan\left(\dfrac{x}{3}\right) \qquad \text{ANSWER A}$$

> **Example 3**
>
> $$\lim_{x\to\infty}\left(\frac{3x^2+x-2}{x^2-x+1}\right) =$$
>
> **A** 0 **B** 1 **C** 2 **D** 3 **E** ∞

In this example it is best to rearrange the function by dividing the numerator by the denominator to give

$$\frac{3x^2+x-2}{x^2-x+1} = 3+\frac{4x-5}{x^2-x+1}$$

Thus $\lim_{x\to\infty}\left(\frac{3x^2+x-2}{x^2-x+1}\right) = \lim_{x\to\infty}\left(3+\frac{4x-5}{x^2-x+1}\right) = 3,$

since the term $\dfrac{4x-5}{x^2-x+1}\to 0$ as $x\to\infty$. $\qquad$ ANSWER D

Alternatively, by dividing the numerator and denominator by x^2

$$\lim_{x\to\infty}\left(\frac{3x^2+x-2}{x^2-x+1}\right) = \lim_{x\to\infty}\left(\frac{3+\dfrac{1}{x}-\dfrac{2}{x^2}}{1-\dfrac{1}{x}+\dfrac{1}{x^2}}\right) = 3 \quad \text{ANSWER D}$$

Note Remember that it may be possible to find a limit by direct substitution, e.g.

$$\lim_{x\to 1}\left(\frac{x^2-3x+4}{x^2+2}\right) = \tfrac{2}{3}$$

However direct substitution may produce an indeterminate value such as $\dfrac{0}{0}$ or $\dfrac{\infty}{\infty}$. In such cases some way of modifying the function must be found as in Example 3.

Another useful technique is to factorize the expression and cancel common factors, e.g.

$$\lim_{x\to 1}\left(\frac{x^2-5x+4}{x^2-3x+2}\right) = \lim_{x\to 1}\frac{(x-4)(x-1)}{(x-2)(x-1)} = \lim_{x\to 1}\left(\frac{x-4}{x-2}\right) = 3$$

De l'Hopital's rule can also be used to evaluate the limit of $\dfrac{f(x)}{g(x)}$ as $x\to a$ if $f(a)$ and $g(a)$ are both zero.

The rule states that $\lim\limits_{x\to a}\left(\dfrac{f(x)}{g(x)}\right) = \lim\limits_{x\to a}\left(\dfrac{f'(x)}{g'(x)}\right)$ provided that the second limit exists, e.g.

$$\lim_{x\to 4}\left(\frac{x^2-16}{x-4}\right) = \lim_{x\to 4}\left(\frac{2x}{1}\right) = 8.$$

The graph of $y = \dfrac{x^2-16}{x-4}$ is, in fact, the straight line $y = x+4$ except for $x = 4$, i.e., a straight line with a break in it.

Multiple choice questions (Type B) Answer according to the table

A	B	C	D	E
1, 2, 3	1, 3	2, 3	2	3
correct	only	only	only	only

Example 4

If $y = \cosh x - \sinh x$

1 $\dfrac{dy}{dx} = y$ **2** $\dfrac{d}{dx}(xy) = y(1-x)$ **3** as $x \to \infty$, $\dfrac{dy}{dx}$ tends to zero

---

Since $y = \cosh x - \sinh x$, $\dfrac{dy}{dx} = \sinh x - \cosh x = -y$ \hfill (1)

Thus statement 1 is incorrect.

$$\frac{d}{dx}(xy) = x\frac{dy}{dx}+y \qquad \text{by the product rule}$$

$$= -xy+y \qquad \text{since } \frac{dy}{dx} = -y \text{ from (1)}$$

$$\frac{d}{dx}(xy) = y(1-x)$$

Thus statement 2 is correct.

Writing $\cosh x = \tfrac{1}{2}(e^x+e^{-x})$ and $\sinh x = \tfrac{1}{2}(e^x-e^{-x})$ we have
$$y = \cosh x - \sinh x = e^{-x}$$
As $x \to \infty$, $e^{-x} \to 0$ and thus statement 3 is correct.

Hence only statements 2 and 3 are true. \hfill **ANSWER C**

Note The discerning student could tackle this problem by using the fact that $y = e^{-x}$ from the outset.

$$\boxed{\begin{array}{l}
\textbf{Example 5}\\[1em]
\text{If } y = \dfrac{x}{1+x}\\[1em]
\textbf{1} \ \dfrac{dy}{dx} = \dfrac{1}{(1+x)^2} \qquad \textbf{2} \ \dfrac{d}{dx}(\ln y) = \dfrac{1+x}{x}\\[1em]
\textbf{3} \ \dfrac{d}{dx}(\tan^{-1}y) = \dfrac{1}{1+2x+2x^2}
\end{array}}$$

Using the quotient rule $\qquad \dfrac{dy}{dx} = \dfrac{v\dfrac{du}{dx} - u\dfrac{dv}{dx}}{v^2}$

where $u = x$ and $v = 1+x$,

$$\frac{dy}{dx} = \frac{(1+x)\cdot 1 - (x)\cdot 1}{(1+x)^2} = \frac{1}{(1+x)^2} \qquad\qquad (1)$$

Statement 1 is correct.

Let $u = \ln y$ where $y = \dfrac{x}{1+x}$

Hence $\dfrac{du}{dy} = \dfrac{1}{y}$ and $\dfrac{dy}{dx} = \dfrac{1}{(1+x)^2}$ $\qquad\qquad$ from (1)

Using the function of a function rule

$$\frac{du}{dx} = \frac{du}{dy} \times \frac{dy}{dx}$$

$$\frac{du}{dx} = \frac{1}{y} \times \frac{1}{(1+x)^2} = \frac{(1+x)}{x} \times \frac{1}{(1+x)^2} = \frac{1}{x(1+x)}$$

Thus statement 2 is false.

Let $z = \tan^{-1}y$

$\tan z = y$ and $\dfrac{d}{dx}(\tan z) = \dfrac{dy}{dx}$; $\qquad \sec^2 z\,\dfrac{dz}{dx} = \dfrac{dy}{dx}$

$$\sec^2 z\,\frac{dz}{dx} = \frac{1}{(1+x)^2} \qquad\qquad \text{from (1)}$$

$$\frac{dz}{dx} = \frac{1}{(1+x)^2}\left(\frac{1}{\sec^2 z}\right) = \frac{1}{(1+x)^2}\left(\frac{1}{1+\tan^2 z}\right)$$

$$\frac{dz}{dx} = \frac{1}{(1+x)^2}\left[\frac{1}{1+\dfrac{x^2}{(1+x)^2}}\right] = \frac{1}{(1+x)^2+x^2}$$

$$\frac{dz}{dx} = \frac{d}{dx}(\tan^{-1}y) = \frac{1}{1+2x+2x^2}$$

Thus statement 3 is correct.

Hence only statements 1 and 3 are correct. $\qquad$ **ANSWER B**

Note It is very important to take special note of the variables when differentiating. In statement 2 for example

$$\frac{d}{dy}(\ln y) = \frac{1}{y} = \frac{1+x}{x} \quad \text{but} \quad \frac{d}{dx}(\ln y) = \frac{1}{y}\frac{dy}{dx}$$

Short questions

In these questions the answer has to be evaluated but attention is usually confined to one particular topic on the syllabus. The questions will test a basic method or an application of differentiation to fairly straightforward problems.

Example 6

Differentiate the function $x^2 - 2x - 3$ with respect to x from first principles.

Consider the curve $y = x^2 - 2x - 3$ as shown in Figure 9. Let $P(x, y)$ be a general point of the curve and $Q(x+\delta x, y+\delta y)$ a neighbouring point.

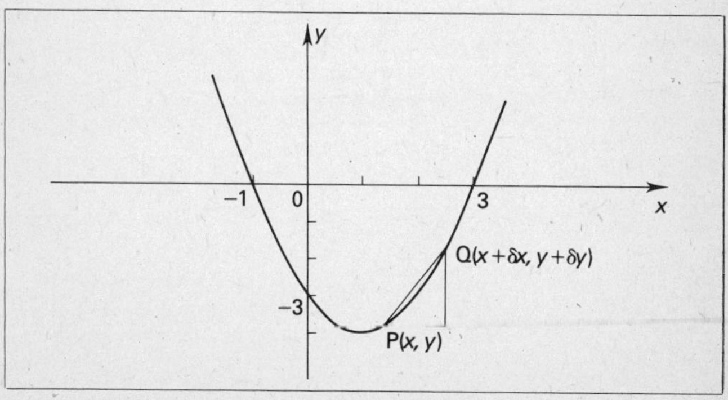

Figure 9

Since Q lies on the curve, the coordinates $(x+\delta x, y+\delta y)$ satisfy the equation of the curve.

$$y+\delta y = (x+\delta x)^2 - 2(x+\delta x) - 3 \tag{1}$$

Also, since $P(x, y)$ lies on the curve

$$y = x^2 - 2x - 3 \tag{2}$$

Subtracting (2) from (1)

$$\begin{aligned}
\delta y &= (x+\delta x)^2 - 2(x+\delta x) - 3 - (x^2 - 2x - 3) \\
&= x^2 + 2x(\delta x) + (\delta x)^2 - 2x - 2(\delta x) - 3 - x^2 + 2x + 3 \\
&= 2x(\delta x) - 2(\delta x) + (\delta x)^2
\end{aligned}$$

The gradient of the chord $PQ = \dfrac{\delta y}{\delta x} = \dfrac{2x(\delta x) - 2(\delta x) + (\delta x)^2}{\delta x}$

$$= 2x - 2 + (\delta x)$$

Now the gradient of the curve at a point is defined as the gradient of the tangent at that point. This is the limit of the gradient of the chord PQ as $\delta x \to 0$.

Hence $\dfrac{dy}{dx} = \lim_{\delta x \to 0} \left(\dfrac{\delta y}{\delta x}\right) = \lim_{\delta x \to 0} [2x - 2 + (\delta x)] = 2x - 2$

Example 7
Differentiate

    (a) $(\ln x)^x$ (b) $\dfrac{(1+x^2)(2-x)^3}{(1+x^3)^2}$ with respect to x.

(a) Let $y = (\ln x)^x$

Taking logarithms of both sides we have

$$\ln y = \ln (\ln x)^x = x \ln (\ln x)$$

Differentiating with respect to x

$$\frac{1}{y}\frac{dy}{dx} = x \frac{d}{dx}[\ln (\ln x)] + \ln (\ln x)$$

$$= x\left[\frac{1}{\ln x} \times \frac{1}{x}\right] + \ln (\ln x) = \frac{1}{\ln x} + \ln (\ln x)$$

$$\therefore \quad \frac{dy}{dx} = y\left[\frac{1}{\ln x} + \ln (\ln x)\right] = (\ln x)^x\left[\frac{1}{\ln x} + \ln (\ln x)\right]$$

$$\frac{dy}{dx} = (\ln x)^{x-1} + (\ln x)^x \ln (\ln x)$$

(b) The idea of taking logarithms can also be used for a function of this form instead of the product and quotient rules.

Let $y = \dfrac{(1+x^2)(2-x)^3}{(1+x^3)^2}$

Taking logarithms of both sides we have

$$\ln y = \ln(1+x^2) + \ln(2-x)^3 - \ln(1+x^3)^2$$

$$\Rightarrow \qquad \ln y = \ln(1+x^2) + 3\ln(2-x) - 2\ln(1+x^3)$$

Differentiating with respect to x

$$\frac{1}{y}\frac{dy}{dx} = \frac{2x}{(1+x^2)} - \frac{3}{(2-x)} - \frac{6x^2}{(1+x^3)}$$

$$\frac{1}{y}\frac{dy}{dx} = \frac{2x(2-x)(1+x^3) - 3(1+x^2)(1+x^3) - 6x^2(1+x^2)(2-x)}{(1+x^2)(2-x)(1+x^3)}$$

$$\frac{1}{y}\frac{dy}{dx} = \frac{x^5 - 8x^4 + 3x^3 - 17x^2 + x^2 + 4x - 3}{(1+x^2)(2-x)(1+x^3)}$$

$$\therefore \quad \frac{dy}{dx} = \frac{(1+x^2)(2-x)^3}{(1+x^3)^2}\left[\frac{x^5 - 8x^4 + 3x^3 - 17x^2 + 4x - 3}{(1+x^2)(2-x)(1+x^3)}\right]$$

$$\therefore \quad \frac{dy}{dx} = \frac{(2-x)^2}{(1+x^3)^3}(x^5 - 8x^4 + 3x^3 - 17x^2 + 4x - 3)$$

Note Remember that logarithms are taken to base e (written $\ln x$) in order that $\dfrac{d}{dx}(\ln x) = \dfrac{1}{x}$. These are called natural logarithms.

Example 8

If $y = \dfrac{2\cos nx}{x}$ where n is a constant prove that

$$x\frac{d^2y}{dx^2} + 2\frac{dy}{dx} + n^2xy = 0.$$

Since $y = \dfrac{2\cos nx}{x}$ it follows that $xy = 2\cos nx$.

Differentiating implicitly with respect to x

$$x\frac{dy}{dx} + y = -2n\sin nx$$

Differentiating implicitly a second time with respect to x

$$x\frac{d^2y}{dx^2}+\frac{dy}{dx}+\frac{dy}{dx}=-2n^2\cos nx$$

$$x\frac{d^2y}{dx^2}+2\frac{dy}{dx}=-n^2xy \quad \text{or} \quad x\frac{d^2y}{dx^2}+2\frac{dy}{dx}+n^2xy=0$$

Note This solution illustrates the power of implicit differentiation since the alternative method of forming the first and second differentials separately is more lengthy and cumbersome.

Long questions

The solutions of these questions usually require an understanding of more than one part of the syllabus.

In Example 8 it was necessary to differentiate twice but sometimes further differentials may be required. This can be achieved by noticing a pattern in successive differentials or by using Leibniz's theorem.

Example 9

If $y=\tan^{-1}x$ prove that $\dfrac{dy}{dx}=\dfrac{1}{1+x^2}$ and

$$(1+x^2)\frac{d^2y}{dx^2}+2x\frac{dy}{dx}=0.$$

Use Leibniz's theorem to show that if $y_n=\dfrac{d^ny}{dx^n}$

$$(1+x^2)y_{n+2}+2(n+1)xy_{n+1}+n(n+1)y_n=0$$

Hence find the first four terms of the expansion of $\tan^{-1}x$.

If $y=\tan^{-1}x$ then $\tan y=x$.

Differentiating with respect to x $\qquad \sec^2y\dfrac{dy}{dx}=1$

Hence $\dfrac{dy}{dx}=\dfrac{1}{\sec^2y}=\dfrac{1}{1+\tan^2y}=\dfrac{1}{1+x^2}$ $\qquad\qquad$ (1)

This can be written in the form $(1+x^2)\dfrac{dy}{dx}=1$

Differentiating implicitly with respect to x

$$(1+x^2)\frac{d^2y}{dx^2}+2x\frac{dy}{dx}=0 \tag{2}$$

If $\dfrac{dy}{dx},\dfrac{d^2y}{dx^2},\ldots,\dfrac{d^ry}{dx^r},\ldots$ are denoted by $y_1,y_2,\ldots,y_r,\ldots$ respectively, Equation (2) becomes

$$(1+x^2)y_2+2xy_1=0$$

Differentiating n times with respect to x using Leibniz's theorem on each of the products we have,

$$(1+x^2)y_{n+2}+{}^nC_1 2xy_{n+1}+{}^nC_2(2)y_n+2xy_{n+1}+{}^nC_1(2)y_n=0$$
$$(1+x^2)y_{n+2}+2nxy_{n+1}+n(n-1)y_n+2xy_{n+1}+2ny_n=0$$
$$(1+x^2)y_{n+2}+2(n+1)xy_{n+1}+n(n+1)y_n=0 \tag{3}$$

It is now possible to evaluate the value of the various differentials. Putting $x=0$ in Equations (1), (2) and (3) we obtain,

from (1) $\qquad\qquad\qquad y_1(0)=1$

from (2) $\qquad\qquad\qquad y_2(0)=0$

from (3) $\qquad\qquad\quad y_{n+2}(0)=-n(n+1)y_n(0)$

$\therefore\ y_3(0)=(-1)(2)y_1(0)=-2$

$\therefore\ y_4(0)=(-2)(3)y_2(0)=0$

$\therefore\ y_5(0)=(-3)(4)y_3(0)=(-3)(4)(-2)=24$

$\therefore\ y_6(0)=(-4)(5)y_4(0)=(-4)(5)(0)=0$

$\therefore\ y_7(0)=(-5)(6)y_5(0)=(-5)(6)(24)=-720$

Using Maclaurin's theorem, i.e.

$$f(x)=f(0)+xf'(0)+\frac{x^2}{2!}f''(0)+\frac{x^3}{3!}f'''(0)+\ldots$$

$$\tan^{-1}(x)=0+x(1)+\frac{x^3}{3!}(-2)+\frac{x^5}{5!}(24)+\frac{x^7}{7!}(-720)+\ldots$$

$$\tan^{-1}(x)=x-\frac{x^3}{3}+\frac{x^5}{5}-\frac{x^7}{7}+\ldots$$

Example 10

Find the coordinates of any turning points or points of inflexion on the curve $y=x^4-4x^3$. Sketch the curve.

The gradient of the curve is given by $\dfrac{dy}{dx}=4x^3-12x^2$.

Turning points occur when $\dfrac{dy}{dx} = 0$, i.e.

when
$$4x^3 - 12x^2 = 0$$
$$4x^2(x-3) = 0$$
$$x = 0 \ \text{or} \ x = 3$$

If $x = 0$, $y = 0$.
If $x = 3$, $y = (3)^4 - 4(3)^3 = -27$.

Thus the gradient of the curve is zero at the points $(0,0)$ and $(3, -27)$. The nature of the turning points can be determined by considering the second differential.

$$\frac{d^2y}{dx^2} = 12x^2 - 24x = 12x(x-2)$$

When $x = 0$, $\dfrac{d^2y}{dx^2} = 0$ but $\dfrac{d^3y}{dx^3} = 24x - 24 \neq 0$.

Thus the point $(0,0)$ is a point of inflexion.

When $x = 3$, $\dfrac{d^2y}{dx^2} > 0$.

Thus the point $(3, -27)$ is a minimum point.
The curve has a point of inflexion at $(0,0)$ and a minimum at $(3, -27)$.
Also, since when $y = 0$, $x = 0$ or $x = 4$, the curve cuts the x-axis at the points $(0,0)$ and $(4,0)$.
As $x \to \pm \infty$, $y \to \infty$.
The curve is sketched in Figure 10.

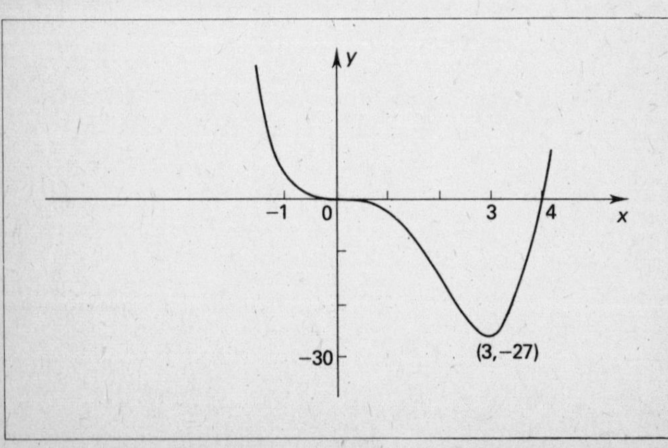

Figure 10

Note
1. Remember that a point of inflexion does not necessarily require $\dfrac{dy}{dx} = 0$, although it is in this case.

2. The nature of the turning points can be determined by considering the gradient of the curve on either side of the point.

x	$x < 0$	0	$x > 0$
$\dfrac{dy}{dx}$	$-$ ve	0	$-$ ve

x	$x < 3$	3	$x > 3$
$\dfrac{dy}{dx}$	$-$ ve	0	$+$ ve

Clearly this indicates a point of inflexion at $(0,0)$ and a minimum at $(3, -27)$.

Example 11
A right circular cone is inscribed in a solid sphere of radius R. If x is the distance of the base of the cone from the centre of the sphere, prove that the volume of the cone is given by $V = \frac{1}{3}\pi(R^3 + R^2x - Rx^2 - x^3)$.

Hence find the height of the cone when the volume of the cone is a maximum. Find also, the maximum volume of the cone.

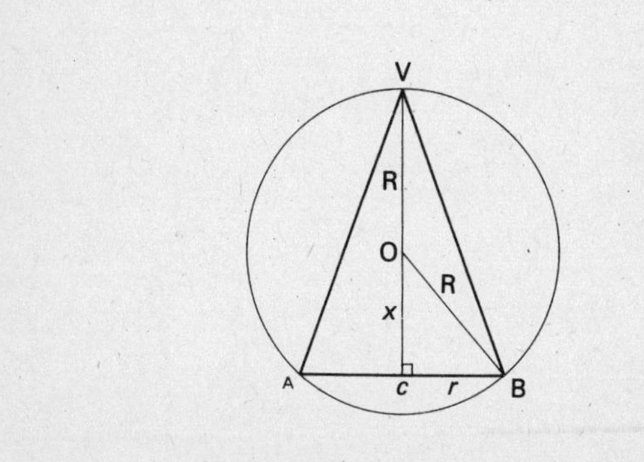

Figure 11

Figure 11 shows a cross-section of the cone VAB inscribed in a sphere, centre O and radius R.

Since $OV = OB = R$ (radius of the sphere) and $OC = x$, it follows from Pythagoras' theorem that

$$CB^2 = OB^2 - OC^2$$
$$= R^2 - x^2$$

$\therefore$ the base radius of the cone, r, is given by

$$r = \sqrt{R^2 - x^2}$$

The volume of a cone $= \frac{1}{3}\pi r^2 h$ where r is the base radius and h is the perpendicular height.

Thus the volume of the cone, V, is given by

$$V = \tfrac{1}{3}\pi(R^2 - x^2)(R + x)$$
$$= \tfrac{1}{3}\pi(R^3 + R^2 x - Rx^2 - x^3)$$

For a maximum volume $\dfrac{dV}{dx} = 0$ and $\dfrac{d^2 V}{dx^2} < 0$

Since $$V = \tfrac{1}{3}\pi(R^3 + R^2 x - Rx^2 - x^3)$$

$$\dfrac{dV}{dx} = \tfrac{1}{3}\pi(R^2 - 2Rx - 3x^2) \quad (R \text{ is constant})$$

When $\dfrac{dV}{dx} = 0$
$$\tfrac{1}{3}\pi(R^2 - 2Rx - 3x^2) = 0$$
$$R^2 - 2Rx - 3x^2 = 0$$
$$(R - 3x)(R + x) = 0$$
$$\therefore \quad x = \tfrac{1}{3}R \ \text{ or } \ x = -R$$

Since $0 < x < R$ in this problem we can discard the solution $x = -R$.

Now $$\dfrac{d^2 V}{dx^2} = \tfrac{1}{3}\pi(-2R - 6x)$$

Thus if $x = \tfrac{1}{3}R, \dfrac{d^2 V}{dx^2} < 0$ which defines a maximum.

The maximum volume of the cone occurs when $x = \tfrac{1}{3}R$. The height of the cone when $x = \tfrac{1}{3}R$ is $\tfrac{4}{3}R$.

$\therefore$ the maximum volume of the cone is

$$V = \tfrac{1}{3}\pi\left(R^2 - \frac{R^2}{9}\right)\left(R + \frac{R}{3}\right) = \tfrac{1}{3}\pi\left(\frac{8R^2}{9}\right)\left(\frac{4R}{3}\right) = \frac{32\pi R^3}{81}$$

Note

1. The solution $x = -R$ does, in fact, give the minimum value that V would take if it was considered simply as a function of x without the physical limitations of the problem.

Note that when $x = -R$, $\dfrac{d^2 V}{dx^2} > 0$.

2. Remember that the expression for the volume, V, must only contain one variable and if this is not immediately the case then steps must be taken to eliminate the extra variables before differentiating, i.e., in this example the volume of the cone, V, is given by

$$V = \tfrac{1}{3}\pi r^2 h = \tfrac{1}{3}\pi r^2 (R+x)$$

The variable, r, was eliminated by using the result $r^2 = R^2 - x^2$ to produce the expression $V = \tfrac{1}{3}\pi(R^2 - x^2)(R+x)$ which only contains the one variable, x.

Chapter 4

Integration

Integration can be considered as the reverse of differentiation, i.e., in integration a function is determined from its gradient or as a limit of the summation of small parts. As the former of these two ideas gives an indefinite answer it is called indefinite integration, e.g.,

$$\text{If } y = f(x) \text{ and } f'(x) = x^2 \text{ then } y = \tfrac{1}{3}x^3 + c$$

where c is an arbitrary constant which can only be determined if further boundary conditions are given.

However, the process of adding a large number of small parts to form a sum, as in calculating an area, gives rise to a definite numerical result and is called definite integration.

There are a great number of special methods for integrating a variety of functions but most syllabuses will expect a knowledge of inspection methods, substitution, integration by parts and reduction formulae. These will be applied to algebraic, trigonometrical, exponential and hyperbolic functions.

Applications are varied but generally include finding areas, volumes and centroids of plane figures or solids, mean values, lengths of arcs, areas of sectors, surfaces of revolution and problems involving velocity and acceleration.

Infinite integrals and numerical step by step methods are often required.

Multiple choice questions (Type A) Select the correct answer

Example 1

$$\int \sin^2 x \, dx =$$

A $\tfrac{1}{3}\sin^3 x + c$ **B** $2\sin x \cos x + c$ **C** $\tfrac{1}{2}x + \tfrac{1}{4}\cos 2x + c$
D $\tfrac{1}{2}x - \tfrac{1}{4}\sin 2x + c$ **E** $\tfrac{1}{2}x + \tfrac{1}{2}\sin 2x + c$

Since the powers of trigonometrical functions cannot be integrated immediately it is necessary to transform the integrand into a trigonometrical function of multiple angles.

Since $\sin^2 x = \frac{1}{2}(1 - \cos 2x)$ (Double angle formula)

$$\int \sin^2 x \, dx = \int \tfrac{1}{2} - \tfrac{1}{2}\cos 2x \, dx$$

$$= \tfrac{1}{2}x - \tfrac{1}{4}\sin 2x + c \qquad \text{ANSWER D}$$

Example 2

Figure 12 shows the curve $y = \ln x$. The volume of the solid of revolution formed when the shaded area is rotated about the line $x = 1$ is given by

A $\pi \displaystyle\int_1^2 (1 - \ln x)\,dx$ **B** $\pi \displaystyle\int_1^2 (1 - \ln x)^2 \, dx$

C $\pi \displaystyle\int_0^{\ln 2} (e^y - 1)^2 \, dy$ **D** $\pi \displaystyle\int_0^{\ln 2} (\ln x)^2 \, dy$ **E** $\pi \displaystyle\int_0^{\ln 2} e^{2y} \, dy$

If the shaded area is to be rotated about the line $x = 1$ then an element of volume can be formed by rotating an element of area about the line $x = 1$ and summing.

The volume of the approximate cylinder parallel to the x-axis is $\pi(x - 1)^2 \delta y$.

These elements of volume can be summed between $y = 0$ and $y = \ln 2$ to give, as $\delta y \to 0$,

total volume $= \displaystyle\lim_{y \to 0} \sum_{y=0}^{\ln 2} \pi(x - 1)^2 \delta y$

$= \pi \displaystyle\int_0^{\ln 2} (x - 1)^2 \, dy$

$= \pi \displaystyle\int_0^{\ln 2} (e^y - 1)^2 \, dy \qquad \text{ANSWER C}$

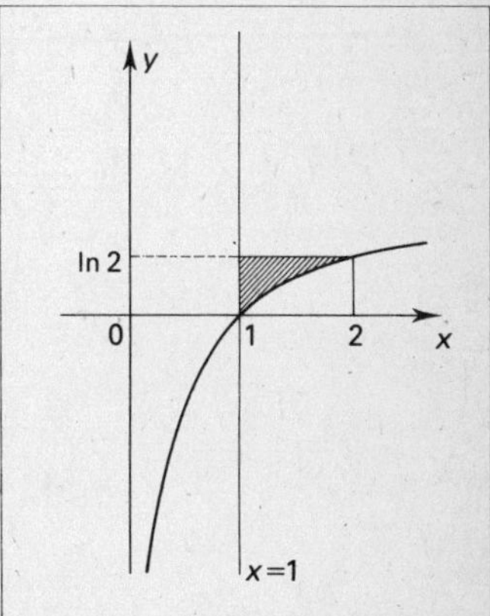

Figure 12

Example 3

$$\int e^{2x}\cos x\, dx =$$

A $\frac{1}{5}e^{2x}(2\cos x+\sin x)+c$ **B** $\frac{1}{2}e^{2x}(\cos x+\sin x)+c$

C $\frac{1}{2}e^{2x}(\cos x-\sin x)+c$ **D** $\frac{1}{3}e^{2x}(2\cos x-\sin x)+c$

E $\frac{2}{5}e^{2x}(\cos x-2\sin x)+c$

Since the integrand is a product we can use integration by parts.

Let $I = \displaystyle\int e^{2x}\cos x\, dx$

Let $u = \cos x \Rightarrow \dfrac{du}{dx} = \sin x$ and $\dfrac{dv}{dx} = e^{2x} \Rightarrow v = \frac{1}{2}e^{2x}$

Since $\displaystyle\int u\,\dfrac{dv}{dx}\,dx = uv - \int v\,\dfrac{du}{dx}\,dx$ we have,

$$I = \int e^{2x}\cos x\, dx = \frac{1}{2}e^{2x}\cos x - \int (-\sin x)\cdot \frac{1}{2}e^{2x}\, dx$$

$$= \frac{1}{2}e^{2x}\cos x + \frac{1}{2}\int e^{2x}\sin x\, dx$$

Repeating this process on the integral produced we have,

Let $u = \sin x \Rightarrow \dfrac{du}{dx} = \cos x$ and $\dfrac{dv}{dx} = e^{2x} \Rightarrow v = \frac{1}{2}e^{2x}$

$\therefore \quad I = \frac{1}{2}e^{2x}\cos x + \frac{1}{2}\left[\frac{1}{2}e^{2x}\sin x - \int \cos x \cdot (\frac{1}{2}e^{2x})\, dx\right]$

$\therefore \quad I = \frac{1}{2}e^{2x}\cos x + \frac{1}{4}e^{2x}\sin x - \frac{1}{4}I$

$\therefore \quad \frac{5}{4}I = \frac{1}{4}e^{2x}(2\cos x + \sin x)$

or $\quad I = \displaystyle\int e^{2x}\cos x\, dx = \frac{1}{5}e^{2x}(2\cos x + \sin x)+c \qquad$ **ANSWER A**

Note that the constant of integration is chosen to be zero when evaluating v. The incorrect answers often given in this problem stem from errors in differentiating $\cos x$ and $\sin x$ and in integrating e^{2x}. Care must be taken with the signs.

Example 4

$$\int_0^1 3^x \, dx =$$

A $\dfrac{2}{\ln 3}$ **B** 2   **C** $2\ln 3$ **D** $-\dfrac{2}{\ln 3}$ **E** no finite value

Let $3^x = u$

Taking logarithms, to base e, of both sides

$$\ln(3^x) = \ln u \Rightarrow x \ln 3 = \ln u$$

Differentiating with respect to x

$$\ln 3 = \left(\frac{1}{u}\right)\frac{du}{dx} \Rightarrow \frac{dx}{du} = \frac{1}{u \ln 3}$$

Using the method of substitution

$$\int_0^1 3^x \, dx = \int_1^3 3^x \frac{dx}{du} \, du = \int_1^3 u \left(\frac{1}{u}\right)\left(\frac{1}{\ln 3}\right) du$$

$$\therefore \quad \int_0^1 3^x \, dx = \int_1^3 \frac{1}{\ln 3} \, du = \left[\frac{u}{\ln 3}\right]_1^3 = \frac{1}{\ln 3}(3^1 - 3^0)$$

$$\int_0^1 3^x \, dx = \frac{2}{\ln 3} \hspace{3cm} \textbf{ANSWER A}$$

Example 5

The area of one loop of the curve $r^2 = a^2 \cos 2\theta$ is

A a^2 **B** $\tfrac{1}{4}a^2$ **C** $\tfrac{1}{2}a^2$ **D** $2a^2$ **E** $\dfrac{a^2}{2\sqrt{2}}$

The general result giving the area of a sector contained between the lines $\theta = \alpha$ and $\theta = \beta$ is given by $\int_\alpha^\beta \tfrac{1}{2}r^2 \, d\theta$. Since $\cos 2\theta$ is unaltered when θ is replaced by $-\theta$ or by $(\pi - \theta)$ the curve is symmetrical about the lines $\theta = 0$ and $\theta = \tfrac{1}{2}\pi$. It consists of two loops touching the lines $\theta = \pm\tfrac{1}{4}\pi$ at the pole, as shown in Figure 13.

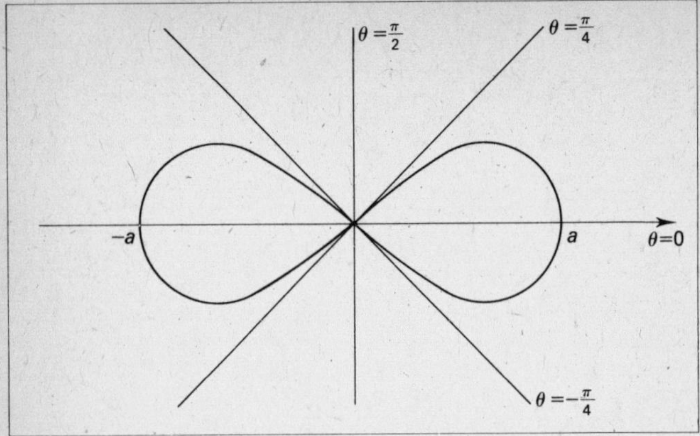

Figure 13

Thus the limits will be $\theta = -\frac{1}{4}\pi$ and $\theta = \frac{1}{4}\pi$.

$$\therefore \quad \text{Area} = \int_{-\pi/4}^{\pi/4} \tfrac{1}{2}r^2 \, d\theta = 2 \int_{0}^{\pi/4} \tfrac{1}{2}r^2 \, d\theta$$

$$= \int_{0}^{\pi/4} a^2 \cos 2\theta \, d\theta = [\tfrac{1}{2}a^2 \sin 2\theta]_0^{\pi/4} = \tfrac{1}{2}a^2$$

ANSWER C

Multiple choice questions (Type B) Answer according to the table

A	B	C	D	E
1, 2, 3	1, 3	2, 3	2	3
correct	only	only	only	only

Example 6

If $f(x) = \dfrac{1}{\sqrt{1+x^2}}$

1 $\displaystyle\int_{0}^{1} f(x)\,dx = \sinh^{-1}1$

2 $f(x)$ is symmetrical about the y-axis.

3 $f'(x)$ at $x = 0$ is zero.

Now $\displaystyle\int_0^1 f(x)\,dx = \int_0^1 \frac{1}{\sqrt{1+x^2}}\,dx$

Using the substitution $x = \sinh u \Rightarrow \dfrac{dx}{du} = \cosh u.$

For this transformation $x = 0 \Rightarrow u = 0$ and $x = 1 \Rightarrow u = \sinh^{-1} 1.$

$$\therefore \quad \int_0^1 f(x)\,dx = \int_{x=0}^{x=1} f(x)\frac{dx}{du}\,du$$

$$= \int_0^{\sinh^{-1}1} \frac{1}{\sqrt{1+\sinh^2 u}}\cosh u\,du$$

$$\therefore \quad \int_0^1 f(x)\,dx = \int_0^{\sinh^{-1}1} \frac{1}{\cosh u}\cosh u\,du = \int_0^{\sinh^{-1}1} \cdot\,du$$

$$= [u]_0^{\sinh^{-1}1} = \sinh^{-1}1$$

Statement 1 is correct.

Since $f(x)$ is an even function (i.e., $f(x) = f(-x)$) the graph is symmetrical about the y-axis.
Statement 2 is correct.

$$\text{If } f(x) = \frac{1}{\sqrt{1+x^2}} \text{ then } f'(x) = \frac{d}{dx}\left[\frac{1}{(1+x^2)^{1/2}}\right]$$

Using the quotient rule for differentiation

$$\therefore \quad f'(x) = \frac{(1+x^2)^{1/2}\cdot 0 - 1\cdot\frac{1}{2}(1+x^2)^{-1/2}\cdot 2x}{(1+x^2)} = -\frac{x}{(1+x^2)^{3/2}}$$

Thus $f'(x) = 0$ when $x = 0$ and statement 3 is correct. Statements 1, 2 and 3 are all true.   ANSWER A

Example 7
If $f(x) = \sin x$ and $g(x) = \cos x$

1   $\displaystyle\int \frac{f(x)}{g(x)}\,dx = \ln(k\cos x)$ where k is a constant.

2   $\displaystyle\int_0^{\pi/2} f(x)\,dx = \int_0^{\pi/2} g(x)\,dx$

3   $\displaystyle\int_0^{\pi/2} \ln f(x)\,dx = \int_0^{\pi/2} \ln g(x)\,dx$

Since $\dfrac{f(x)}{g(x)} = \dfrac{\sin x}{\cos x} = \tan x$ we have

$$\int \tan x \, dx = \int \frac{\sin x}{\cos x} \, dx = -\int \frac{-\sin x}{\cos x} \, dx = -\ln(k\cos x)$$

Statement 1 is incorrect.

$$\text{Now} \quad \int_0^{\pi/2} f(x)\,dx = \int_0^{\pi/2} \sin x \, dx = [-\cos x]_0^{\pi/2}$$

$$= 0 - (-1) = 1$$

$$\text{Also} \quad \int_0^{\pi/2} g(x)\,dx = \int_0^{\pi/2} \cos x \, dx = [\sin x]_0^{\pi/2} = 1 - 0 = 1$$

Statement 2 is correct.

Let $x = -\dfrac{\pi}{2} - u \Rightarrow \dfrac{dx}{du} = -1$

Under this transformation, $x = 0 \Rightarrow u = \dfrac{\pi}{2}$ and $x = \dfrac{\pi}{2} \Rightarrow u = 0$.

$$\text{Hence} \quad \int_0^{\pi/2} \ln f(x)\,dx = \int_{x=0}^{x=\pi/2} \ln(\sin x) \frac{dx}{du}\,du$$

$$= \int_{\pi/2}^0 \ln\left[\sin\left(\tfrac{1}{2}\pi - u\right)\right](-1)\,du$$

$$= -\int_{\pi/2}^0 \ln(\cos u)\,du = \int_0^{\pi/2} \ln(\cos u)\,du$$

Replacing u by x we obtain

$$\int_0^{\pi/2} \ln(\sin x)\,dx = \int_0^{\pi/2} \ln(\cos x)\,dx$$

Statement 3 is correct.

Statements 2 and 3 only are true. $\hfill$ ANSWER C

Example 8

If $I = \displaystyle\int_0^\infty e^{-3x}\,dx$

1 The x-axis is an asymptote to the curve $y = e^{-3x}$.
2 I is infinite   **3** $I = \tfrac{1}{3}$

As $x \to \infty$, $e^{-3x} \to 0$ and thus the curve approaches the x-axis from above. The line $y = 0$ is an asymptote to the curve.
Statement 1 is correct.

Although the range of integration is infinite the area tends to a limiting value and we define

$$\int_0^\infty f(x)\,dx = \lim_{N \to \infty} \int_0^N f(x)\,dx$$

$$\therefore \quad \int_0^\infty e^{-3x}\,dx = \lim_{N \to \infty} \int_0^N e^{-3x}\,dx = \lim_{N \to \infty} \left[-\tfrac{1}{3}e^{-3x} \right]_0^N$$

$$= \lim_{N \to \infty} \left[\tfrac{1}{3} - \tfrac{1}{3}e^{-3N} \right] = \tfrac{1}{3}$$

Thus statement 2 is incorrect since the integral is finite and statement 3 is correct.

Statements 1 and 3 only are true. **ANSWER B**

Example 9

$$\text{If } I = \int_0^1 \frac{dx}{\sqrt{1-x^2}}$$

1 I is infinite.   **2** The integral cannot be evaluated.
3 $I = \tfrac{1}{2}\pi$.

In this integral the integrand is infinite at the upper limit $(x = 1)$ and hence we consider its value when the upper limit is $1 - \alpha$ where α is a small positive constant. The value of I will be given by the limit as $\alpha \to 0$ assuming that the limit exists.

$$\therefore \quad \lim_{\alpha \to 0} \int_0^{1-\alpha} \frac{dx}{\sqrt{1-x^2}} = \lim_{\alpha \to 0} \left[\sin^{-1} x \right]_0^{1-\alpha}$$

$$= \lim_{\alpha \to 0} \left[\sin^{-1}(1-\alpha) \right]$$

As $\alpha \to 0$, $\sin^{-1}(1-\alpha) \to \sin^{-1} 1 = \tfrac{1}{2}\pi$, (taking the principal value).

Hence $I = \tfrac{1}{2}\pi$, and statement 3 is correct. Statements 1 and 2 are thus incorrect.

Statement 3 only is true. **ANSWER E**

Multiple choice questions (Type C) For each question two statements are given. Answer

A if 1 implies 2 and 2 implies 1
B if 2 implies 1 but 1 does not imply 2
C if 1 implies 2 but 2 does not imply 1
D if 1 denies 2 and 2 denies 1
E if none of these relations holds

Example 10

1 $\dfrac{dy}{dx} = \dfrac{2x-1}{x^2+4}$ **2** $y = \ln(x^2+4) - \tfrac{1}{2}\tan^{-1}(\tfrac{1}{2}x)$

Now

$$\int \frac{2x-1}{x^2+4}\,dx = \int \frac{2x}{x^2+4}\,dx - \int \frac{1}{x^2+4}\,dx$$

$$= \ln(x^2+4) - \tfrac{1}{2}\tan^{-1}(\tfrac{1}{2}x) + c$$

Hence statement 2 gives a particular integral of statement 1 and although statement 2 implies statement 1, statement 1 does not imply statement 2. **ANSWER B**

Short questions
These questions test a small part of the syllabus and require a basic integration method or an application of integration.

Example 11
Find the smallest positive value of n for which

$$\int_0^n 4\cos 3x \cos x \, dx = 0.$$

Using the factor formula

$$\cos P + \cos Q = 2\cos\tfrac{1}{2}(P+Q)\cos\tfrac{1}{2}(P-Q)$$
$$\tfrac{1}{2}(P+Q) = 3x \Rightarrow P+Q = 6x \tag{1}$$
$$\tfrac{1}{2}(P-Q) = \ x \Rightarrow P-Q = 2x \tag{2}$$

Adding and subtracting equations (1) and (2) we have
$$2P = 8x \ \text{ and } \ 2Q = 4x \Rightarrow P = 4x \ \text{ and } \ Q = 2x$$

$$\therefore \quad \int_0^n 4\cos 3x \cos x \, dx = \int_0^n 2\cos 4x + 2\cos 2x \, dx$$

$$= [\tfrac{1}{2}\sin 4x + \sin 2x]_0^n = \tfrac{1}{2}\sin 4n + \sin 2n = 0$$
$$\therefore \quad \sin 2n \cos 2n + \sin 2n = \sin 2n(\cos 2n + 1) = 0$$
$$\therefore \quad \sin 2n = 0 \ \text{ or } \ \cos 2n = -1$$

If $\sin 2n = 0$ $2n = 0, \pi, 2\pi, \ldots$ $n = 0, \frac{1}{2}\pi, \pi, \ldots$
If $\cos 2n = -1$ $2n = \pi, 3\pi, \ldots$ $n = \frac{1}{2}\pi, \frac{3}{2}\pi, \ldots$
Thus the smallest value of $n > 0$ is $\frac{1}{2}\pi$.

Example 12

Evaluate $\displaystyle\int_3^4 \frac{x+4}{x^2-x-2}\, dx$

Since the denominator will factorize we use the method of partial fractions to evaluate this integral.

Let

$$\frac{x+4}{x^2-x-2} \equiv \frac{x+4}{(x-2)(x+1)} \equiv \frac{A}{(x-2)} + \frac{B}{(x+1)}$$

$$\equiv \frac{A(x+1)+B(x-2)}{(x-2)(x+1)}$$

Equating the numerators

$$x+4 \equiv A(x+1)+B(x-2)$$

Let $x = 2 \Rightarrow 6 = 3A \Rightarrow A = 2$
Let $x = -1 \Rightarrow 3 = -3B \Rightarrow B = -1$

$$\therefore \int_3^4 \frac{x+4}{x^2-x-2}\, dx = \int_3^4 \frac{2}{(x-2)} - \frac{1}{(x+1)}\, dx$$

$$= [2\ln(x-2) - \ln(x+1)]_3^4$$
$$= 2\ln 2 - \ln 5 + \ln 4 = \ln \tfrac{16}{5}$$

Example 13

Evaluate $\displaystyle\int \frac{dx}{x^2+6x+10}$

In this example the denominator does not factorize so we complete the square.

$$\therefore \int \frac{dx}{x^2+6x+10} = \int \frac{dx}{(x+3)^2+1} = \tan^{-1}(x+3)+c$$

Example 14

Evaluate $\displaystyle\int_0^{\pi/2} \frac{1}{1+\sin x}\, dx$

Use the substitution $t = \tan\frac{1}{2}x$. From the double angle formula for $\tan 2A$ we have

$$\tan x = \frac{2t}{1-t^2} \Rightarrow \cos x = \frac{1-t^2}{1+t^2} \quad \text{and} \quad \sin x = \frac{2t}{1+t^2}$$

Also differentiating with respect to x

$$\frac{dt}{dx} = \tfrac{1}{2}\sec^2\frac{x}{2} = \tfrac{1}{2}\left(1+\tan^2\frac{x}{2}\right) = \tfrac{1}{2}(1+t^2) \Rightarrow \frac{dx}{dt} = \frac{2}{1+t^2}$$

Hence

$$\int_0^{\pi/2} \frac{1}{1+\sin x}\,dx = \int_0^1 \frac{1}{\left(1+\dfrac{2t}{1+t^2}\right)}\left(\frac{2}{(1+t^2)}\right)dt$$

$$= \int_0^1 \frac{2}{1+2t+t^2}\,dt = \int_0^1 \frac{2}{(1+t)^2}\,dt = \left[\frac{-2}{(1+t)}\right]_0^1$$

$$= -\frac{2}{2} - \left(-\frac{2}{1}\right) = 1$$

Example 15

Evaluate $\int x\tan^{-1}x\,dx$

Integration by parts is a powerful method for integrating a product.

Let $u = \tan^{-1}x \Rightarrow \dfrac{du}{dx} = \dfrac{1}{1+x^2}$ and $\dfrac{dv}{dx} = x \Rightarrow v = \tfrac{1}{2}x^2$

Using the result $\displaystyle\int u\frac{dv}{dx}\,dx = uv - \int v\frac{du}{dx}\,dx$ we obtain

$$\int x\tan^{-1}x\,dx = \tfrac{1}{2}x^2\tan^{-1}x - \int \tfrac{1}{2}x^2\left[\frac{1}{(1+x^2)}\right]dx$$

$$= \tfrac{1}{2}x^2\tan^{-1}x - \tfrac{1}{2}\int \frac{x^2}{(1+x^2)}\,dx$$

$$= \tfrac{1}{2}x^2\tan^{-1}x - \tfrac{1}{2}\int \left(1 - \frac{1}{(1+x^2)}\right)dx$$

$$= \tfrac{1}{2}x^2\tan^{-1}x - \tfrac{1}{2}x + \tfrac{1}{2}\tan^{-1}x + c$$

This can be extended to a more general case which gives rise to a reduction formula expressing one integral in terms of another.

Example 16

If $I_n = \displaystyle\int_0^2 x^n e^{ax} \, dx$ obtain a reduction formula relating I_n and I_{n-1}. Hence evaluate $\displaystyle\int_0^2 x^2 e^{4x} \, dx$.

Let $u = x^n \Rightarrow \dfrac{du}{dx} = nx^{n-1}$ and $\dfrac{dv}{dx} = e^{ax} \Rightarrow v = \dfrac{1}{a}e^{ax}$

Integrating by parts

$$I_n = \int_0^2 x^n e^{ax} \, dx = \left[\frac{1}{a}x^n e^{ax}\right]_0^2 - \frac{n}{a}\int_0^2 x^{n-1} e^{ax} \, dx$$

Now $I_{n-1} = \displaystyle\int_0^2 x^{n-1} e^{ax} \, dx$ since it is of the same form as I_n except that n is replaced by $n-1$.

$$\therefore \; I_n = \left[\frac{x^n}{a}e^{ax}\right]_0^2 - \frac{n}{a}I_{n-1} \quad \text{or} \quad I_n = \frac{2^n e^{2a}}{a} - \frac{n}{a}I_{n-1} \tag{1}$$

Now $I_2 = \displaystyle\int_0^2 x^2 e^{4x} \, dx$ where $a = 4$

Thus using (1)
$$I_2 = \frac{2^2 e^8}{4} - \frac{2}{4}\int_0^2 x e^{4x} \, dx$$

$$\therefore \; I_2 = e^8 - \tfrac{1}{2}I_1 = e^8 - \tfrac{1}{2}\left[\frac{2e^8}{4} - \tfrac{1}{4}I_0\right] = \tfrac{3}{4}e^8 + \tfrac{1}{8}I_0$$

$$= \tfrac{3}{4}e^8 + \tfrac{1}{8}\int_0^2 e^{4x} \, dx$$

$$= \tfrac{3}{4}e^8 + \left[\tfrac{1}{32}e^{4x}\right]_0^2 = \tfrac{3}{4}e^8 + \tfrac{1}{32}e^8 - \tfrac{1}{32}$$

$$= \tfrac{1}{32}(25e^8 - 1)$$

Long questions

Full questions may be set on the various applications of integration and may consist of a complete problem or involve several different parts.

Example 17

Find the mean value of sinh x for $0 \leqslant x \leqslant \ln 2$. Find also the perimeter of the region defined by the inequalities $0 \leqslant x \leqslant \ln 2$ and $0 \leqslant y \leqslant \sinh x$.

If this region is rotated through 2π radians about the x-axis calculate the curved surface area of the solid of revolution so formed.

The mean value of a function over the range $a \leqslant x \leqslant b$ is given by $\dfrac{1}{b-a} \displaystyle\int_a^b f(x)\,dx$.

$\therefore$ the mean value of $\sinh x$ is $\dfrac{1}{\ln 2} \displaystyle\int_0^{\ln 2} \sinh x\,dx$

$$= \frac{1}{\ln 2}[\cosh x]_0^{\ln 2} = \frac{1}{2\ln 2}[e^x + e^{-x}]_0^{\ln 2}$$

$$= \frac{1}{2\ln 2}[(e^{\ln 2} + e^{-\ln 2}) - (e^0 + e^0)]$$

$$= \frac{1}{2\ln 2}(2 + \tfrac{1}{2} - 2) = \frac{1}{4\ln 2} \tag{1}$$

The region defined by $0 \leqslant x \leqslant \ln 2$ and $0 \leqslant y \leqslant \sinh x$ is shown in Figure 14.

Perimeter $= \text{arc}\,OA + OB + AB$
$= \text{arc}\,OA + \tfrac{1}{2}(e^{\ln 2} - e^{-\ln 2}) + \ln 2$

Now the length of the arc OA

is $\displaystyle\int_0^{\ln 2} \sqrt{1 + \left(\frac{dy}{dx}\right)^2}\,dx$

$\text{Arc}\,OA = \displaystyle\int_0^{\ln 2} \sqrt{1 + \cosh^2 x}\,dx$

$\phantom{\text{Arc}\,OA} = \displaystyle\int_0^{\ln 2} \sinh x\,dx = \tfrac{1}{4}$

using the working in (1).

Therefore the total perimeter of the region

$= \tfrac{1}{4} + \tfrac{1}{2}(2 - \tfrac{1}{2}) + \ln 2 = 1 + \ln 2$

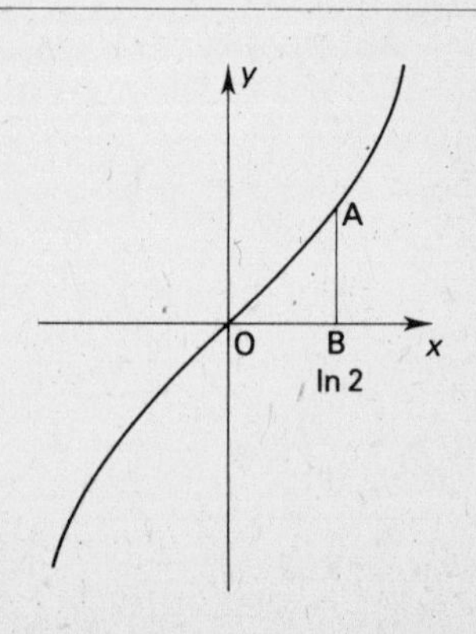

Figure 14

Now when the area is rotated about the x-axis the area of the curved surface of revolution is $\int_0^{\ln 2} 2\pi y \, ds.$

$$\therefore \quad \text{area} = \int_0^{\ln 2} 2\pi y \frac{ds}{dx} dx = \int_0^{\ln 2} 2\pi y \sqrt{1+\left(\frac{dy}{dx}\right)^2} \, dx$$

$$= 2\pi \int_0^{\ln 2} \sinh x \sqrt{1+\cosh^2 x} \, dx = 2\pi \int_0^{\ln 2} \sinh^2 x \, dx$$

$$= 2\pi \int_0^{\ln 2} (\tfrac{1}{2}\cosh 2x - \tfrac{1}{2}) \, dx = \pi[\tfrac{1}{2}\sinh 2x - x]_0^{\ln 2}$$

$$= \pi[\tfrac{1}{4}(e^{2\ln 2} - e^{-2\ln 2}) - \ln 2] = \pi(\tfrac{15}{16} - \ln 2)$$

$\therefore$ the curved surface area is $\pi(\tfrac{15}{16} - \ln 2)$.

Example 18

Find the volume of the solid formed when the region enclosed by the curve $y = \cos x$, the axes and the line $x = \pi/6$ is rotated completely about the x-axis.

Show also that Simpson's rule using 5 ordinates can be used to give a good estimate of the volume.

Consider an element of volume formed by rotating an element of area about the x-axis (see Figure 15).

If $P(x, y)$ and $Q(x + \delta x, y + \delta y)$ are neighbouring points on the curve then the volume formed is $\simeq \pi y^2 \delta x.$

Summing all such elements between $x = 0$ and $x = \pi/6$

$$\text{Total volume} \simeq \sum_{x=0}^{x=\pi/6} \pi y^2 \delta x.$$

In the limit as $\delta x \to 0$

$$\text{volume} = \lim_{\delta x \to 0} \sum_{x=0}^{x=\pi/6} \pi y^2 \delta x$$

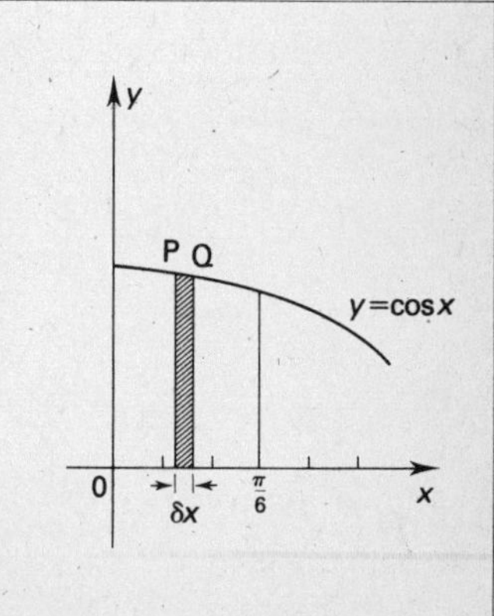

Figure 15

$$= \pi \int_0^{\pi/6} y^2 \, dx = \pi \int_0^{\pi/6} \cos^2 x \, dx = \tfrac{1}{2}\pi \int_0^{\pi/6} (1 + \cos 2x) \, dx$$

$$= \tfrac{1}{2}\pi[x + \tfrac{1}{2}\sin 2x]_0^{\pi/6} = \tfrac{1}{2}\pi(\pi/6 + \tfrac{1}{2}\sin \pi/3) = \pi^2/12 + \sqrt{3}\pi/8$$

The volume of revolution is 1.503 to three decimal places.

Using Simpson's rule for 5 ordinates $(y_1, y_2, y_3, y_4, y_5)$ we have

$$\text{Volume} = \pi\left[\tfrac{1}{3}h(y_1 + 4y_2 + 2y_3 + 4y_4 + y_5)\right]$$

Since the interval is 0 to $\pi/6$, $h = \pi/24$.

As $y^2 = \cos^2 x$

x	0	$\pi/24$	$\pi/12$	$\pi/8$	$\pi/6$
y^2	1	0.9830	0.9330	0.8536	0.7500

$$
\begin{array}{lll}
y_1 = 1.0000 & y_3 = 0.9330 & y_2 = 0.9830 \\
y_5 = 0.7500 & \quad\times\quad 2 & y_4 = 0.8536 \\
\hline
1.7500 & 1.8660 & 1.8366 \\
1.8660 & & \times\quad 4 \\
\hline
7.3464 & & 7.3464 \\
\hline
10.9624 & &
\end{array}
$$

$\therefore$ volume of revolution $= \pi\left[\dfrac{\pi}{72} \times 10.9624\right] = 1.503$ to three decimal places.

Chapter 5

Exponential and Logarithmic Functions

Multiple choice questions (Type A) Select the correct answer

Example 1

$\log_a b =$

A $\dfrac{\log_b c}{\log_a c}$ **B** $\dfrac{\log_a c}{\log_c b}$ **C** $\dfrac{\log_c a}{\log_c b}$ **D** $\dfrac{-\log_c b}{\log_a c}$ **E** $\dfrac{\log_c b}{\log_c a}$

$$L = \log_a b \Rightarrow a^L = b \qquad \text{using logarithm property}$$
$$\Rightarrow \log_c a^L = \log_c b \qquad \text{taking logs to base } c$$
$$\Rightarrow L \log_c a = \log_c b \qquad \text{using logarithm property}$$
$$\Rightarrow L = \frac{\log_c b}{\log_c a} \qquad \text{ANSWER E}$$

Example 2

$\dfrac{d}{dx}(\ln \tan x) =$

A $\dfrac{1}{\tan x}$ **B** $\dfrac{1}{\sin x} - \dfrac{1}{\cos x}$ **C** $\dfrac{\sec^2 x}{\tan x}$ **D** $-\dfrac{\sec x}{\tan x}$ **E** $\dfrac{\cot x}{\tan x}$

Put $u = \tan x$, so $y = \ln \tan x = \ln u$ and $\dfrac{dy}{du} = \dfrac{1}{u}$ and $\dfrac{du}{dx} = \sec^2 x$

$$\frac{dy}{dx} = \frac{dy}{du} \times \frac{du}{dx} = \frac{1}{u} \times \sec^2 x = \frac{\sec^2 x}{\tan x} \qquad \text{ANSWER C}$$

Alternatively, $y = \ln \tan x = \ln(\sin x/\cos x) = \ln \sin x - \ln \cos x$

$$\frac{dy}{dx} = \frac{\cos x}{\sin x} + \frac{\sin x}{\cos x} = \frac{\cos^2 x + \sin^2 x}{\sin x \cos x} = \frac{1}{\sin x \cos x} = \frac{\sec^2 x}{\tan x}$$

Example 3

$$\frac{d}{dx}(\log_{10}x) =$$

A $\dfrac{1}{x}$ **B** $\dfrac{1}{x}\log_{10}e$ **C** $\dfrac{1}{x}\log_e 10$ **D** $\dfrac{1}{10x}$ **E** $\dfrac{10}{x}$

Use the change of base in Question 3 if you cannot remember the result.

$$y = \log_{10}x = \frac{\log_e x}{\log_e 10} = \frac{\ln x}{\ln 10} \Rightarrow \frac{dy}{dx} = \frac{1}{x\ln 10}$$

This does not appear as one of the alternative answers.

However $\quad \ln 10 = \log_e 10 = \dfrac{\log_{10}10}{\log_{10}e} = \dfrac{1}{\log_{10}e}$

so $\quad \dfrac{1}{x\ln 10} = \dfrac{1}{x}\log_{10}e \qquad \dfrac{dy}{dx} = \dfrac{1}{x}\times\dfrac{1}{\log_e 10} = \dfrac{1}{x}\log_{10}e$

ANSWER B

Example 4

The coefficient of x^3 in the expansion of e^{2x} is

A $\frac{1}{6}$ **B** $-\frac{1}{6}$ **C** $\frac{4}{3}$ **D** $-\frac{4}{3}$ **E** 8

$$e^{2x} = 1 + 2x + \frac{(2x)^2}{2!} + \frac{(2x)^3}{3!} + \dots$$

The coefficient of x^3 is $\dfrac{2^3}{3!} = \dfrac{8}{6} = \dfrac{4}{3}$ **ANSWER C**

Example 5

In the expansion of $\ln(1-2x)$ the first two non-zero terms are

A $-2x+2x^2$ **B** $-2x-2x^2$ **C** $2x-4x^2$ **D** $2x+4x^2$
E $-2x-x^2$

$$\ln(1+x) = x - \tfrac{1}{2}x^2 + \tfrac{1}{3}x^3 - \dots$$

$$\Rightarrow \ln(1-2x) = -2x - \tfrac{1}{2}\cdot 4x^2 - \dots = -2x - 2x^2 \qquad \text{**ANSWER B**}$$

> **Example 6**
> For the relation $y = ax^n$ a straight line is obtained by plotting
>
> **A** y against $\log x$ **B** $\log y$ against x **C** y against x
> **D** $\log y$ against $\log x$ **E** none of these

$$y = ax^n \Rightarrow \log y = \log ax^n = \log a + \log x^n = \log a + n \log x$$

Plot $\log y$ on the y-axis and $\log x$ on the x-axis to give $y = nX + \log a$ which is a straight line, of gradient n and y intercept $\log a$.   **ANSWER D**

From the graph the numerical values of n and a are calculated.

> **Example 7**
> Which of the following functions is even?
>
> **A** e^x **B** e^{-x} **C** $\ln x$ **D** $e^x + e^{-x}$ **E** $e^x - e^{-x}$

An even function satisfies $f(x) = f(-x)$ and is symmetrical about the y-axis.

$e^x \neq e^{-x}$ which rules out answers A and B; $\ln(-x)$ is not defined.

$$f(x) = e^x + e^{-x} \Rightarrow f(-x) = e^{-x} + e^x = f(x)$$ **ANSWER D**

The function in answer D is $2\cosh x$, which is even.
The function in answer E is $2\sinh x$, which is odd, i.e.,
$f(-x) = -f(x)$.

> **Example 8**
> The sum to n terms of the series $\ln e + \ln e^2 + \ldots + \ln e^n$ is
>
> **A** $\dfrac{e^n - 1}{e - 1}$ **B** $\dfrac{1}{e - 1}$ **C** $\dfrac{1}{\ln e - 1}$ **D** $\dfrac{\ln e^n - 1}{\ln e - 1}$ **E** $\dfrac{n(n+1)}{2}$

$$\ln e + \ln e^2 + \ln e^3 + \ldots + \ln e^n = 1 + 2 + 3 + \ldots + n = \tfrac{1}{2}n(n+1)$$
                    ANSWER E

> **Example 9**
> $\log_2 32 \div \log_2 4 =$
>
> **A** 4   **B** $2\frac{1}{2}$ **C** 28   **D** 3   **E** 2^3

$$\log_2 32 \div \log_2 4 = \log_2 2^5 \div \log_2 2^2 = 5\log_2 2 \div 2\log_2 2 = 5 \div 2 = 2\tfrac{1}{2}$$

ANSWER B

Multiple choice questions (Type B) Answer according to the table

A	B	C	D	E
1, 2, 3	1, 3	2, 3	2	3
correct	only	only	only	only

Example 10

$$\frac{d}{dx}(\ln x^3) =$$

1 $\dfrac{1}{x^3}$ **2** $\dfrac{3}{x}$ **3** $\dfrac{3x^2}{x^3}$

$$\ln x^3 = 3\ln x \Rightarrow \frac{d}{dx}(\ln x^3) = \frac{3}{x} = \frac{3x^2}{x^3}$$

Answers 2 and 3 are correct. ANSWER C

Example 11

$$I = \int_2^p \frac{1}{x}\,dx = k - \ln 2$$

1 $p = 4 \Rightarrow I = \ln 2$ **2** $p = 1 \Rightarrow k = 0$ **3** $p = e \Rightarrow k = 1$

$$p = 4 \Rightarrow I = \int_2^p \frac{1}{x}\,dx = \ln 4 - \ln 2 = \ln 2$$

statement 1 is correct

$$p = 1 \Rightarrow k = \ln 1 = 0$$

statement 2 is correct

$$p = e \Rightarrow k = \ln e = 1$$

statement 3 is correct

ANSWER A

Example 12

$$\frac{d}{dx}\{\ln(\sec x + \tan x)\} =$$

1 $\sec x$ **2** $\dfrac{1}{\sec x + \tan x}$ **3** $\dfrac{\sec x \tan x + \sec^2 x}{\sec x + \tan x}$

$$\frac{d}{dx}\ln(\sec x + \tan x) = \frac{\sec x \tan x + \sec^2 x}{\sec x + \tan x} \qquad \text{statement 3 is correct}$$

$$= \frac{\sec x(\tan x + \sec x)}{\sec x + \tan x} \qquad \text{statement 1 is correct,}$$
$$\text{statement 2 is wrong}$$
$$= \sec x \qquad \text{ANSWER B}$$

Example 13

For the graph of $x^2 e^{-x}$

1 $x = 2$ gives a maximum point **2** $y \to 0$ as $x \to -\infty$
3 $y \to 0$ as $x \to +\infty$

$$y = x^2 e^{-x} \Rightarrow \frac{dy}{dx} = 2x e^{-x} - x^2 e^{-x} = x(2-x)e^{-x} = 0$$

when $x = 0, 2$.

$$\frac{d^2 y}{dx^2} = 2e^{-x} - 4x e^{-x} + x^2 e^{-x} = (x^2 - 4x + 2)e^{-x} < 0$$

when $x = 2 \Rightarrow$ maximum.

Statement 1 is correct.

$$\text{As } x \to \infty, e^x > x^2 \Rightarrow \frac{x^2}{e^x} \to 0, \text{ i.e., } x^2 e^{-x} \to 0$$

Statement 3 is correct.

As $x \to -\infty$, e^{-x} and x^2 are both large and positive

Statement 2 is wrong.
As $x \to \infty$, e^x gets larger more quickly than x^2

$$\Rightarrow \frac{x^2}{e^x} = 0, \text{ i.e., } x^2 - e^{-x} = 0$$

Statement 3 is correct.   **ANSWER B**

Multiple choice questions (Type C) For each question two
statements are given. Answer

A if 1 implies 2 and 2 implies 1
B if 2 implies 1 but 1 does not imply 2
C if 1 implies 2 but 2 does not imply 1
D if 1 denies 2 and 2 denies 1
E if none of these relations holds

Example 14

1 $\log_a b = c$ **2** $a^c = b$

Statements 1 and 2 are two ways of saying the same thing: they define what a logarithm is. It is easier to remember this definition with a numerical example, i.e.,

$$2^3 = 8 \Leftrightarrow \log_2 8 = 3 \quad \text{or} \quad 10^3 = 1000 \Leftrightarrow \log_{10} 1000 = 3$$

ANSWER A

Example 15

1 $\dfrac{d^2 y}{dx} - 5\dfrac{dy}{dx} + 6y = 0$ **2** $y = 5e^{2x} + 6e^{3x}$

$$y = 5e^{2x} + 6e^{3x} \Rightarrow \frac{dy}{dx} = 10e^{2x} + 18e^{3x} \Rightarrow \frac{d^2 y}{dx^2} = 20e^{2x} + 54e^{3x}$$

$$y'' + 5y' + 6y = 20e^{2x} + 54e^{3x} - 50e^{2x} - 90e^{3x} + 30e^{2x} + 36e^{3x} = 0$$

So statement 2 implies statement 1. y'' and y' are accepted notations for the second and third derivatives.

$y = Ae^{2x} + Be^{3x}$ also satisfies statement 1, so statement 1 does not imply statement 2.

ANSWER B

Example 16

x is a real number

1 $3 = 4\sinh x$ **2** $x = \ln 2$

Method 1 $\sinh x = \frac{3}{4} \Rightarrow x = \sinh^{-1}\frac{3}{4}$

$$\sinh^{-1} x = \ln\left(x + \sqrt{x^2 + 1}\right) = \ln\left(\tfrac{3}{4} + \sqrt{\tfrac{9}{16} + 1}\right) = \ln\left(\tfrac{3}{4} + \sqrt{\tfrac{25}{16}}\right) = \ln 2$$

Statement 1 implies statement 2 and each step is reversible, so statement 2 implies statement 1 and vice versa.

Method 2 $\sinh x = \frac{1}{2}(e^x - e^{-x}) = \frac{1}{2}\left(e^{\ln 2} - \dfrac{1}{e^{\ln 2}}\right) = \frac{1}{2}(2 - \tfrac{1}{2}) = \frac{3}{4}$

Statement 2 implies statement 1 and each step is reversible so statement 1 implies statement 2.

70

Method 3 $3 = 4\sinh x = 4 \times \tfrac{1}{2}(e^x - e^{-x}) = 2e^x - 2e^{-x}$.

Multiplying through by e^x gives $3e^x - 2e^{2x} + 2 = 0$

$\Rightarrow 2e^{2x} - 3e^x - 2 = 0$ (a quadratic in e^x)

$\Rightarrow (2e^x + 1)(e^x - 2) = 0 \Rightarrow e^x = -\tfrac{1}{2}$ (impossible)   or $e^x = 2$

$$\Rightarrow x = \ln 2$$

Statement 1 implies statement 2 and vice versa.   **ANSWER A**

Multiple choice questions (Type D) Each question consists of a problem followed by four pieces of information. Decide whether the problem can be solved with one of the pieces of information omitted and answer

A if 1 could be omitted   **D** if 4 could be omitted
B if 2 could be omitted   **E** if none can be omitted
C if 3 could be omitted

Example 17

Find $\displaystyle \int_a^b \frac{d}{x}\,dx + \int_b^c \frac{d}{x}\,dx$

1 $a = 5$ **2** $b = 6$ **3** $c = 7$ **4** $d = 8$

$$\int_a^b \frac{d}{x}\,dx + \int_b^c \frac{d}{x}\,dx = \int_a^c \frac{d}{x}\,dx = [d \ln x]_a^c = d \ln \frac{c}{a}$$

Value b is not needed.   **ANSWER B**

Example 18

Find c $\displaystyle \int_{a^n}^{a^{n+1}} \frac{b}{x}\,dx$

1 $a = 2$ **2** $c = 6$ **3** $n = 4$ **4** $b = 5$

$$c \int_{a^n}^{a^{n+1}} \frac{b}{x}\,dx = c[b \ln x]_{a^n}^{a^{n+1}} = bc \ln \frac{a^{n+1}}{a^n} = bc \ln a$$

Value n is not needed.   **ANSWER C**

Example 19

$f(x) = \dfrac{ae^x - bce^{-x}}{de^x + de^{-x}}$ is an odd function

1 $a = 6$ **2** $b = 2$ **3** $c = 3$ **4** $d = 4$

$$f(-x) = \frac{ae^{-x} - bce^{x}}{d(e^{-x} + e^{+x})}$$

This will be equal to $-f(x)$ if $a = bc$, so value 4 ($d = 4$) is not needed. **ANSWER D**

Example 20

Find the value of $x = \dfrac{\log_a b^d}{\log_a c}$

1 $a = 2$ **2** $b = 3$ **3** $c = 4$ **4** $d = 5$

$$x = \frac{\log_a b^d}{\log_a c} = d \times \frac{\log_a b}{\log_a c} = d \log_c b \qquad \text{(change of base)}$$

Value 1 ($a = 2$) is unnecessary. **ANSWER A**

Short questions

Example 21

Solve the equations (a). $e^{\ln x} = 5$ and (b) $\ln e^x = 6$.

The usual way to solve equations involving powers is to take logarithms in order to eliminate the powers.

(a) $\ln(e^{\ln x}) = \ln 5$ (b) $\ln e^x = 6$

      $\ln x \cdot \ln e = \ln 5$ $x \ln e = 6$

      $\ln x = \ln 5$ (since $\ln e = 1$) $x = 6$

        $x = 5$

In fact e^x and $\ln x$ are inverse functions, so when applied successively they produce the identity function, i.e.,

$$e^{\ln x} = x \quad \text{and} \quad \ln e^x = x$$

If you can remember these, the equations are trivial.

Example 22

The product rule for differentiation is

$$\frac{d}{dx}(uv) = u\frac{dv}{dx} + v\frac{du}{dx}$$

Derive the product rule for differentiating a product of 3 functions uvw by using logarithms.

$$f = uvw \Rightarrow \ln f = \ln(uvw) = \ln u + \ln v + \ln w$$

$$\Rightarrow \frac{1}{f}\frac{df}{dx} = \frac{1}{u}\frac{du}{dx} + \frac{1}{v}\frac{dv}{dx} + \frac{1}{w}\frac{dw}{dx}$$

$$\Rightarrow \frac{df}{dx} = \frac{f}{u}\frac{du}{dx} + \frac{f}{v}\frac{dv}{dx} + \frac{f}{w}\frac{dw}{dx}$$

$$\Rightarrow \frac{d}{dx}(uvw) = vw\frac{du}{dx} + uw\frac{dv}{dx} + uv\frac{dw}{dx}$$

Example 23

Find $\displaystyle\int \frac{2x}{x^2-1}\,dx$ (a) directly and (b) by partial fractions.

(a) You must recognize that the numerator is the derivative of the denominator and since $\displaystyle\int \frac{f'(x)}{f(x)}\,dx = \ln f(x)$

$$\int \frac{2x}{x^2-1}\,dx = \ln(x^2-1)+c$$

(b) $\displaystyle\frac{2x}{x-1} = \frac{2x}{(x-1)(x+1)} = \frac{a}{x-1} + \frac{b}{x+1}$

$$\Rightarrow 2x = a(x+1) + b(x-1)$$

By covering up or using $x = 1$ and -1, $a = 1$ and $b = 1$

$$\int \frac{2x}{x-1}\,dx = \int\left(\frac{1}{x-1} + \frac{1}{x+1}\right)dx = \ln(x-1) + \ln(x+1) + c$$

$$= \ln(x^2-1)+c$$

Example 24

Differentiate $\log_{10} x$ from first principles.

$$y = \log_{10} x \Rightarrow y + \delta y = \log_{10}(x+\delta x)$$

$$\Rightarrow \delta y = \log_{10}(x+\delta x) - \log_{10} x = \log_{10}\frac{(x+\delta x)}{x}$$

$$\frac{dy}{dx} = \lim_{\delta x \to 0}\frac{\delta y}{\delta x} = \lim_{\delta x \to 0}\frac{1}{\delta x}\log\left(\frac{x+\delta x}{x}\right) = \lim_{\delta x \to 0}\frac{1}{\delta x}\log\left(1+\frac{\delta x}{x}\right)$$

Write $\dfrac{\delta x}{x} = t \Rightarrow \dfrac{1}{\delta x} = \dfrac{1}{xt}$ and as $\delta x \to 0$ so $t \to 0$.

$$\frac{dy}{dx} = \lim_{t \to 0} \frac{1}{xt} \log(1+t) = \lim_{t \to 0} \frac{1}{x} \log(1+t)^{1/t} \text{ using } \log N^p = p \log N.$$

Now we need to find $\lim_{t \to 0} (1+t)^{1/t} = e = 2.718\,28\ldots$

(see Example 32)

$$\frac{dy}{dx} = \frac{1}{x} \log_{10} e \qquad \text{(which verifies Example 1)}$$

If instead we took logarithms to base e we have

$$y = \log_e x \Rightarrow \frac{dy}{dx} = \frac{1}{x} \log_e e = \frac{1}{x} \text{ and therefore } \int \frac{1}{x} dx = \ln x + c$$

Example 25

Find the Cartesian equations of the curves whose parametric equations are

(a) $x = 3 \cosh t, \ y = 2 \sinh t$ (b) $x = 3 \sec t, \ y = 2 \tan t$

(a) $\cosh^2 t - \sinh^2 t = 1 \Rightarrow \dfrac{x^2}{9} - \dfrac{y^2}{4} = 1$ which is a hyperbola symmetrical about both axes. As t takes real values only one branch of the hyperbola is described.

(b) $\sec^2 t - \tan^2 t = 1 \Rightarrow \dfrac{x^2}{9} - \dfrac{y^2}{4} = 1$, the same hyperbola.

$-\frac{1}{2}\pi < t < \frac{1}{2}\pi$ gives one branch of the hyperbola (x positive).

$\frac{1}{2}\pi < t < 1\frac{1}{2}\pi$ gives the other branch (x negative).

Example 26

Solve $a^b = b^a$ for $a > b > 0$ where a and b are positive integers

$$a^b = b^a \Rightarrow b \log a = a \log b \Rightarrow \frac{\log a}{a} = \frac{\log b}{b}$$

Consider $y = \dfrac{\ln x}{x}; \ \dfrac{dy}{dx} = \dfrac{1 - \ln x}{x^2} = 0$ if $\ln x = 1 \Rightarrow x = e$.

If a solution exists, then $b = 1$ or 2; $b = 1 \Rightarrow a = 1$ (contradiction); $b = 2 \Rightarrow a = 4$ (no other solution).

$$\boxed{\begin{array}{l}\textbf{Example 27}\\[4pt]\text{Obtain the expansion for } \ln(1+x) \text{ in ascending powers of } x\\ \text{and deduce the expansion for } \ln(1-x). \text{ Use these two to}\\ \text{work out } \ln(1-x^2) \text{ and check this by comparing it with}\\ \ln(1-x), \text{ with } x^2 \text{ replacing } x.\end{array}}$$

$$f(x) = f(0) + xf'(0) + \frac{x^2}{2!}f''(0) + \ldots$$

$$f(x) = \ln(1+x) \Rightarrow f'(x) = \frac{1}{1+x}, \; f''(x) = \frac{-1}{(1+x)^2}, \; f'''(x) = \frac{2}{(1+x)^3}$$

$$f''''(x) = \frac{-3!}{(1+x)^4}, \; f^r(x) = \frac{(-1)^{r-1}(r-1)!}{(1+x)^r}$$

$$x = 0 \Rightarrow f(0) = 0, \; f'(0) = 1, \; f''(0) = -1, \; f'''(0) = 2, \; f''''(0) = -3!,$$
etc.

$$\ln(1+x) = x - \frac{x^2}{2} + \frac{2x^3}{3!} - \frac{3!x^4}{4!} + \frac{4!x^5}{5!}$$

$$= x - \frac{x^2}{2} + \frac{x^3}{3} - \frac{x^4}{4} + \frac{x^5}{5} - \ldots + \frac{(-1)^{r-1}x^r}{r} \tag{1}$$

Substituting $-x$ for x gives

$$\ln(1-x) = -x - \tfrac{1}{2}x^2 - \tfrac{1}{3}x^3 - \tfrac{1}{4}x^4 - \ldots \tag{2}$$

Adding (1) and (2) gives $\ln(1+x) + \ln(1-x) = \ln(1+x)(1-x) = \ln(1-x^2)$. Equations (1) and (2) gives $\ln(1-x) = -x^2 - \tfrac{1}{2}x^4 - \tfrac{1}{3}x^6 - \tfrac{1}{4}x^8$ which is just the same as the series obtained by substituting x^2 for x in (2).

$$\boxed{\begin{array}{l}\textbf{Example 28}\\[4pt]\text{Find } \dfrac{d}{dx}\{\tfrac{1}{4}(e^x + e^{-x})^2 - \tfrac{1}{4}(e^x - e^{-x})^2\}\end{array}}$$

Some students may recognize this as

$$\frac{d}{dx}(\cosh^2 x - \sinh^2 x) = \frac{d}{dx}(1) = 0$$

Otherwise the difference of two squares will give

$$\frac{d}{dx}(\tfrac{1}{4}(e^x + e^{-x})^2 - \tfrac{1}{4}(e^x - e^{-x})^2) = \frac{d}{dx}\left(\frac{2e^x}{2} \times \frac{2e^{-x}}{2}\right) = \frac{d}{dx}(1) = 0$$

> **Example 29**
> For $x > 0$, find the minimum point on the curve $y = x^x$.

$$\ln y = \ln x^x = x \ln x \Rightarrow \frac{1}{y}\frac{dy}{dx} = \ln x + x \cdot \frac{1}{x} = 1 + \ln x$$

$$\frac{dy}{dx} = y(1 + \ln x) = x^x(1 + \ln x) = 0 \Rightarrow \ln x = -1 \Rightarrow x = e^{-1} = \frac{1}{e} = 0.37$$

$$x = e^{-1} \Rightarrow y = \left(\frac{1}{e}\right)^{1/e} = 0.69; \quad \frac{d^2y}{dx^2} = x^x(1 + \ln x)^2 + x^x \cdot \frac{1}{x} > 0$$

when $x = 0.37$. Minimum point is $(0.37, 0.69)$.

> **Example 30**
> Define $\sinh x$ and $\cosh x$ and differentiate both. Derive the
> addition formulae for $\sinh(a+b)$, $\cosh(a+b)$, $\sinh 2a$ and
> $\cosh 2a$ using the formulae equivalent to $\sin^2 x + \cos^2 x = 1$.

$$\sinh x = \tfrac{1}{2}(e^x - e^{-x}) \Rightarrow \frac{d}{dx}(\sinh x) = \tfrac{1}{2}(e^x + e^{-x}) = \cosh x$$

$$\cosh x = \tfrac{1}{2}(e^x + e^{-x}) \Rightarrow \frac{d}{dx}(\cosh x) = \tfrac{1}{2}(e^x - e^{-x}) = \sinh x$$

$$\sinh a + \cosh a = e^a \quad \text{and} \quad \cosh a - \sinh a = e^{-a}$$
$$(\cosh a + \sinh a)(\cosh a - \sinh a) = \cosh^2 a - \sinh^2 a = 1$$
$$e^{a+b} = e^a \cdot e^b = (\cosh a + \sinh a)(\cosh b + \sinh b)$$
$$= \cosh a \cdot \cosh b + \cosh a \cdot \sinh b + \sinh a \cdot \cosh b$$
$$+ \sinh a \cdot \sinh b \tag{1}$$
$$e^{-a-b} = e^{-a}e^{-b} = (\cosh a - \sinh a)(\cosh b - \sinh b)$$
$$= \cosh a \cdot \cosh b - \cosh a \cdot \sinh b - \sinh a \cdot \cosh b$$
$$+ \sinh a \cdot \sinh b \tag{2}$$

$(1) + (2)$ gives $e^{a+b} + e^{-a-b} = 2\cosh a \cosh b + 2\sinh a \sinh b$

$$\cosh(a+b) = \tfrac{1}{2}(e^{a+b} + e^{-a-b}) = \cosh a \cosh b + \sinh a \sinh b \tag{3}$$

$(1) - (2)$ gives $\sinh(a+b) = \sinh a \cosh b + \cosh a \sinh b \tag{4}$

Put $b = a$ in (3) to give $\quad \cosh 2a = \cosh^2 a + \sinh^2 a$
$$= 1 + 2\sinh^2 a$$
$$\text{using } \cosh^2 a = 1 + \sinh^2 a$$
$$= 2\cosh^2 a - 1$$
$$\text{as } \sinh^2 a = \cosh^2 a - 1$$

Put $b = a$ in (4) to give $\quad \sinh 2a = 2\sinh a \cosh a$

Example 31

Show that $y = \sinh^{-1}x = \ln\left(x + \sqrt{x^2+1}\right)$

Method 1 $y = \sinh^{-1}x \Rightarrow x = \sinh y$ and $\cosh^2 y = 1 + \sinh^2 y = 1 + x^2$

$\quad e^y = \cosh y + \sinh y = x + \sqrt{x^2+1} \Rightarrow y = \ln\left(x + \sqrt{x^2+1}\right)$

Method 2 $y = \sinh^{-1}x \Rightarrow x = \sinh y = \frac{1}{2}(e^y - e^{-y})$

$$\Rightarrow 2x = e^y - e^{-y} \Rightarrow e^{2y} - 2xe^y - 1 = 0$$

$$\text{(a quadratic in } e^y)$$

$$\Rightarrow e^y = \frac{2x \pm \sqrt{4x^2+4}}{2} = x \pm \sqrt{x^2+1}$$

Since $\sqrt{x^2+1} > x$ for real x, and $e^y > 0$ we discard the negative sign

$$e^y = x \pm \sqrt{x^2+1} \Rightarrow y = \ln\left(x + \sqrt{x^2+1}\right)$$

Osborne's rule provides a way of obtaining hyperbolic identities from the corresponding trigonometric formulae. Replace $-\sin^2 x$ by $+\sinh^2 x$ so that $\cos^2 x + \sin^2 x = 1$ becomes $\cosh^2 x - \sinh^2 x = 1$. Also $\sec^2 x = 1 + \tan^2 x$ becomes $\mathrm{sech}^2 x = 1 - \tanh^2 x$ because there is an implied $\sin^2 x$ contained within $\tan^2 x \left(\tan^2 x = \dfrac{\sin^2 x}{\cos^2 x}\right)$.

Example 32

Use the first few terms of the Binomial expansion of $(1+t)^{1/t}$ to find the value of $e = \lim_{t \to 0}(1+t)^{1/t}$ by substituting $t = 0$ in the expanded series.

Compare your value with the print-out on a calculator ($e = 2.718\,281\,828$)

$$(1+x)^n \simeq 1 + nx + n(n-1)\frac{x^2}{2!} + n(n-1)(n-2)\frac{x^3}{3!} + \ldots \text{ gives}$$

$$(1+t)^{1/t} \simeq 1 + \frac{1}{t}\cdot t + \frac{1}{t}\left(\frac{1}{t}-1\right)\frac{t^2}{2!} + \frac{1}{t}\left(\frac{1}{t}-1\right)\left(\frac{1}{t}-2\right)\frac{t^3}{3!} + \ldots$$

Multiply each bracket by a 't' taken from the powers of 't'.

$$(1+t)^{1/t} \simeq 1 + 1 + \frac{(1-t)}{2} + \frac{(1-t)(1-2t)}{6} + \frac{(1-t)(1-2t)(1-3t)}{24} + \ldots$$

Putting $t = 0$ gives $e = \lim_{t \to 0} (1+t)^{1/t} = 2 + \frac{1}{2} + \frac{1}{6} + \frac{1}{24} + \frac{1}{120} + \ldots$

$e = 2.5 + 0.166 + 0.041\ 66 + 0.008\ 333 + 0.001\ 388\ 8 + 0.000\ 198\ 412$
$ = 2.718\ 253\ 7$

This is correct to 3 decimal places since the last term calculated is still affecting the value of the fourth decimal place. My calculator gives a value of $e = 2.718\ 281\ 828$ which is accurate to 8 decimal places. More terms in the series give greater accuracy.

I am told that e is an irrational number (like π, $\sqrt{2}$, $\sqrt{3}$, etc.) i.e. certainly not a recurring decimal, but my calculator suggests a bank of 4 recurring figures '1828'. Is this correct? Work out the series to the tenth or eleventh decimal place (it does not take long with a calculator). Remember that all your terms must be expressed to 11 places of decimals.

Example 33

Obtain the expansion for e^x in ascending powers of x and use it to find the value of e.

$$f(x) = f(0) + xf'(0) + \frac{x^2}{2!} f''(0) + \frac{x^3}{3!} f'''(0) + \ldots$$

$$f(x) = e^x \Rightarrow f'(x) = e^x = f''(x) = f'''(x) = \ldots$$

$$x = 0 \Rightarrow f'(0) = 1 = f''(0) = f'''(0) = \ldots$$

$$e^x = 1 + x + \frac{x^2}{2!} + \frac{x^3}{3!} + \ldots + \frac{x^r}{r!} + \ldots$$

$$x = 1 \text{ gives } e = 1 + 1 + \frac{1}{2} + \frac{1}{6} + \frac{1}{24} + \ldots = 2.718\ 28 \ldots$$

Chapter 6
Coordinate Geometry

The study of the properties of lines, curves and planes using the methods of algebra and calculus is called coordinate geometry. The depth to which this branch of the subject is covered and the content of the course differs considerably according to the syllabus being followed. It is possible that only the straight line and the circle will be needed for the single subject whereas the double subject will expect knowledge of the conics.

Parameters are usually included and students will be expected to be able to deal with equations such as those of tangents, normals and chords in parametric form.

Multiple choice questions (Type A) Select the correct answer

Example 1
The equation of the straight line in Figure 16 is

A $2y - 3x = 6$ **B** $3y + 2x = 9$ **C** $2y + 3x = 9$
D $2y + 3x = 6$ **E** $3y - 2x = 9$

The gradient of the line is

$$\frac{3-0}{0-2} = -\frac{3}{2}$$

The equation of a line of gradient m passing through the point (x_1, y_1) is

$$y - y_1 = m(x - x_1)$$

This line passes through $(0,3)$ so its equation is

$$y - 3 = -\tfrac{3}{2}(x - 0)$$

$$\Rightarrow 2y + 3x = 6$$

ANSWER D

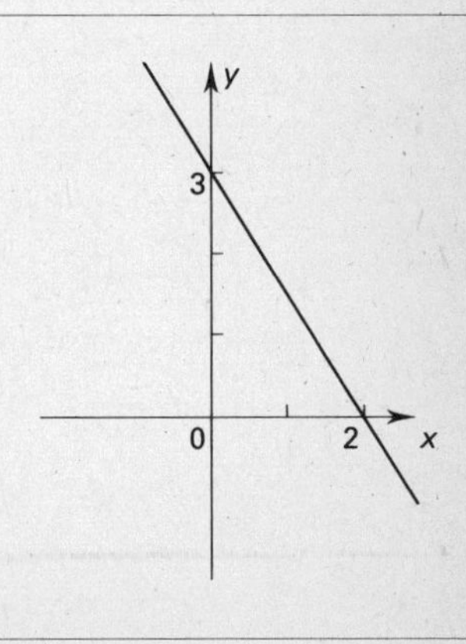

Figure 16

Example 2

The coordinates of the point C which divides the line segment joining the points $A(-3,-4)$ and $B(-1,6)$ internally in the ratio of $3:1$ are

A $(-2\frac{1}{2},-1\frac{1}{2})$ **B** $(-1\frac{1}{2},3\frac{1}{2})$ **C** $(-2\frac{2}{3},-1\frac{1}{3})$
D $(-1\frac{2}{3},2\frac{2}{3})$ **E** $(-1\frac{1}{2},-2\frac{1}{2})$

From Figure 17 it can be seen that triangles ADC and AFB are similar.

Hence
$$\frac{AC}{AB} = \frac{AD}{AF} = \frac{DC}{FB} = \frac{3}{4}$$

$\therefore \quad AD = \tfrac{3}{4}AF = \tfrac{3}{4}\{-1-(-3)\} = \tfrac{3}{4}\times 2 = 1\tfrac{1}{2}$

and $DC = \tfrac{3}{4}FB = \tfrac{3}{4}\{6-(-4)\} = \tfrac{3}{4}\times 10 = 7\tfrac{1}{2}$

Therefore, the coordinates of C are $(-3+1\tfrac{1}{2},-4+7\tfrac{1}{2}) = (-1\tfrac{1}{2},3\tfrac{1}{2})$

ANSWER B

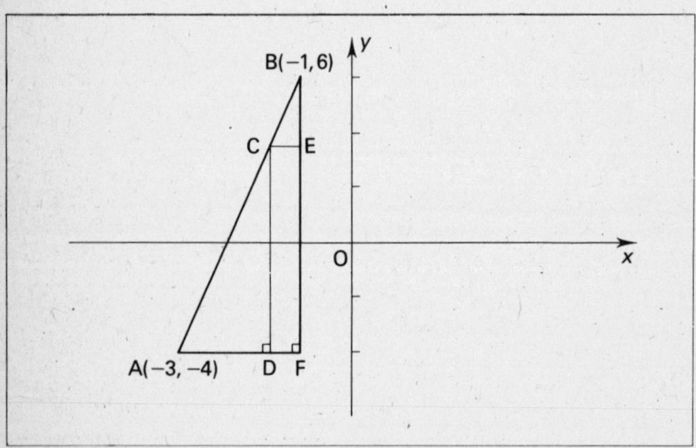

Figure 17

Alternatively, the general result used in coordinate geometry and vector geometry can be used. This states that the coordinates of a point dividing a line internally in the ratio m to n are

$$\left(\frac{nx_1+mx_2}{m+n},\ \frac{ny_1+my_2}{m+n}\right)$$

In this case $m:n = 3:1 \Rightarrow$ coordinates of C are

$$\left(\frac{-3(1)+3(-1)}{4}, \frac{-4(1)+6(3)}{4} \right)$$

$$= (-1\tfrac{1}{2}, 3\tfrac{1}{2})$$

Example 3

The distance of the point $(-1,2)$ from the line $y = \tfrac{1}{2}(6-3x)$ is

A $\dfrac{7}{\sqrt{13}}$ **B** $\dfrac{5}{13}$ **C** $\dfrac{2}{\sqrt{13}}$ **D** $\dfrac{5}{\sqrt{13}}$ **E** $\dfrac{2}{13}$

The distance of the point (x_1, y_1) from the line $ax + by + c = 0$ is given by

$$p = \pm \frac{ax_1 + by_1 + c}{\sqrt{a^2 + b^2}}$$

where the sign is chosen to give a positive distance.

Rearranging the given equation to $3x + 2y - 6 = 0$ we have, in this case,

$$p = \pm \frac{3(-1) + 2(2) - 6}{\sqrt{3^2 + 2^2}} = \frac{5}{\sqrt{13}}$$

(choosing the negative sign)

ANSWER D

Example 4

The equation of the straight line perpendicular to the line $3x + 4y - 2 = 0$ at the point $(6, -4)$ is

A $3y - 4x + 36 = 0$ **B** $3y + 4x - 12 = 0$ **C** $3y - 4x - 34 = 0$
D $3y + 4x - 2 = 0$ **E** $3y - 4x + 12 = 0$

The product of the gradients of two perpendicular lines is -1. The gradient of the line $3x + 4y - 2 = 0$ is $-\tfrac{3}{4}$ and thus the gradient of the perpendicular is $\tfrac{4}{3}$.

Since this line passes through the point $(6, -4)$ its equation is

$$y - (-4) = \tfrac{4}{3}(x - 6) \Rightarrow 3y + 12 = 4x - 24 \Rightarrow 3y - 4x + 36 = 0.$$

ANSWER A

Example 5
The centre of the circle $2x^2 + 2y^2 - 3x + 5y - 8 = 0$ is

A $(-3,5)$ **B** $(-\frac{3}{2},\frac{5}{2})$ **C** $(-\frac{3}{4},\frac{5}{4})$ **D** $(\frac{3}{2},-\frac{5}{2})$ **E** $(\frac{3}{4},-\frac{5}{4})$

The given equation can be written as $x^2 + y^2 - \frac{3}{2}x + \frac{5}{2}y - 4 = 0$.
This gives $(x - \frac{3}{4})^2 + (y + \frac{5}{4})^2 = 4 + \frac{34}{16} = \frac{49}{8}$.
This is the equation of a circle centre $(\frac{3}{4}, -\frac{5}{4})$, radius $\dfrac{7\sqrt{2}}{4}$.

$$\text{ANSWER E}$$

Alternatively, it would be possible to compare the given equation with the standard equation $x^2 + y^2 + 2gx + 2fy + c = 0$ where the centre is $(-g, -f)$.

Clearly, $g = -\frac{3}{4}$ and $f = \frac{5}{4}$ and the centre is $(\frac{3}{4}, -\frac{5}{4})$.

Example 6
The focus of the parabola $y = 4x - x^2$ is at the point

A $(-2, \frac{15}{4})$ **B** $(2, -\frac{17}{4})$ **C** $(2, \frac{15}{4})$ **D** $(-2, \frac{17}{4})$ **E** $(2, -\frac{15}{4})$

The equation $y = 4x - x^2$ can be rearranged to give

$$x^2 - 4x = -y \Rightarrow (x-2)^2 = -y + 4 \Rightarrow (x-2)^2 = -(y-4)$$

Using the substitutions $X = x - 2$ and $Y = -y + 4$ we obtain an equation $X^2 = Y$.
Comparing this with the standard equation $X^2 = 4aY$ whose focus is at $(0,a)$ we have $a = \frac{1}{4}$.
Now if $X = 0$, then $x = 2$ and if $Y = a = \frac{1}{4}$, then $y = \frac{15}{4}$.
The focus of the parabola is at $(2, \frac{15}{4})$.

$$\text{ANSWER C}$$

Example 7
The equation $3x^2 - 10xy + 3y^2$ represents

A an ellipse   **B** a hyperbola   **C** a rectangular hyperbola
D a circle   **E** a line pair

Since the equation $3x^2 - 10xy + 3y^2 = 0$ can be written in the form $(3x - y)(x - 3y) = 0$, it represents a pair of lines: $3x - y = 0$ and $x - 3y = 0$.   ANSWER E

A	B	C	D	E
1, 2, 3	1, 3	2, 3	2	3
correct	only	only	only	only

Example 8
The equation of a circle is $x^2 + y^2 = 8$.

1 Any point on the circle can be represented in terms of a parameter θ by $(2\sqrt{2}\cos\theta, 2\sqrt{2}\sin\theta)$.
2 The equation of the tangent at $(2,2)$ is $x + y = 4$.
3 The line $y = -x - 4$ is a tangent to the circle.

If θ is the parameter then $x = 2\sqrt{2}\cos\theta$ and $y = 2\sqrt{2}\sin\theta$.

Thus $\cos\theta = \dfrac{x}{2\sqrt{2}}$ and $\sin\theta = \dfrac{y}{2\sqrt{2}}$.

Squaring and adding

$$\frac{x^2}{8} + \frac{y^2}{8} = \cos^2\theta + \sin^2\theta = 1$$

$\therefore$ $x^2 + y^2 = 8$ and the point $(2\sqrt{2}\cos\theta, 2\sqrt{2}\sin\theta)$ satisfies the equation for all values of θ.
Statement 1 is correct.

If $x^2 + y^2 = 8$ then, differentiating implicitly with respect to x, we
have $2x + 2y\dfrac{dy}{dx} = 0 \Rightarrow \dfrac{dy}{dx} = -\dfrac{x}{y}$.

At $(2,2)$ the gradient of the tangent is therefore $-\dfrac{2}{2} = -1$.

Hence the equation of the tangent is

$$y - 2 = -1(x - 2) \Rightarrow x + y = 4$$

Statement 2 is correct.
Consider the intersection of the line $y = -x - 4$ and the circle.

Solving simultaneously

$$x^2 + (-x - 4)^2 = 8 \Rightarrow x^2 + x^2 + 8x + 16 = 8$$

$\therefore$ $2x^2 + 8x + 8 = 0 \Rightarrow x^2 + 4x + 4 = 0$ or $(x + 2)^2 = 0$.

Since this produces a repeated root the two cutting points are coincident and the line is a tangent to the circle.

Statement 3 is correct.

Statements 1, 2 and 3 are all true. **ANSWER A**

Example 9

If $P(2t^2, 4t)$ is a point on the parabola $y^2 = 8x$,

1 The tangent at P meets the y-axis at $(0, 2t^2)$.

2 The normal at P passes through the focus for only one real value of t.

3 The normal at P meets the curve again where $y = -\dfrac{(8 + 4t^2)}{t}$.

If $y^2 = 8x$, then, differentiating implicitly with respect to x

$$2y\frac{dy}{dx} = 8 \Rightarrow \frac{dy}{dx} = \frac{4}{y}$$

Hence the gradient of the tangent at $P(2t^2, 4t)$ is $\dfrac{1}{t}$.

The equation of the tangent is

$$y - 4t = \frac{1}{t}(x - 2t^2) \Rightarrow ty = x + 2t^2$$

This line meets the y-axis when $x = 0$, i.e., at the point $(0, 2t)$.

Statement 1 is incorrect.

If the gradient of the tangent is $\dfrac{1}{t}$ then the gradient of the normal is $-t$ and its equation will be

$$y - 4t = -t(x - 2t^2) \Rightarrow y + tx = 4t + 2t^3$$

Since this line passes through the focus $(2, 0)$ we have

$$2t = 4t + 2t^3 \Rightarrow 2t^3 + 2t = 0 \Rightarrow 2t(t^2 + 1) = 0$$

The only real solution is $t = 0$. Statement 2 is correct.

The equation of the normal at P is $y + tx = 4t + 2t^3$. This meets the curve $y^2 = 8x$ when

$$y + \frac{ty^2}{8} = 4t + 2t^3 \quad \text{or} \quad ty^2 + 8y - 16t(2 + t^2) = 0 \qquad (1)$$

Since the lines intersect at $P(2t^2, 4t)$, $y - 4t$ is a factor.

$$\therefore \ ty^2 + 8y - 16t(2 + t^2) = \{y - 4t\}\{ty + (8 + 4t^2)\} = 0$$

Thus the lines intersect again when

$$ty = -(8+4t^2) \quad \text{or} \quad y = -\frac{(8+4t^2)}{t}.$$

Statement 3 is correct.

Statements 2 and 3 only are true. $\hspace{4cm}$ **ANSWER C**

Alternatively, the quadratic equation (1) can be solved by noting that if α and β are the roots of the equation, then

$$\text{sum of the roots} = \alpha + \beta = -\frac{8}{t}$$

But $\beta = 4t$ (known) and hence $\alpha = -4t - \dfrac{8}{t}$ or $\dfrac{-(4t^2+8)}{t}$.

Multiple choice questions (Type C) For each question two statements are given. Answer

A if 1 implies 2 and 2 implies 1
B if 2 implies 1 but 1 does not imply 2
C if 1 implies 2 but 2 does not imply 1
D if 1 denies 2 and 2 denies 1
E if none of these relations holds

Example 10

1 The equation of an ellipse is $\dfrac{x^2}{25} + \dfrac{y^2}{16} = \dfrac{1}{4}$.

2 The foci of an ellipse are $(\pm 1\frac{1}{2}, 0)$.

The equation can be written as $\dfrac{4x^2}{25} + \dfrac{y^2}{4} = 1$ and comparing with

the standard equation of an ellipse, $\dfrac{x^2}{a^2} + \dfrac{y^2}{b^2} = 1$, we have $a = \frac{5}{2}$ and $b = 2$.

If e is the eccentricity then $b^2 = a^2(1-e^2)$.

Hence $\hspace{3cm} 4 = \frac{25}{4}(1-e^2) \Rightarrow e = \frac{3}{5}$

The foci of the standard ellipse are $(\pm ae, 0)$ and thus the foci of the given ellipse are $(\pm\frac{3}{2}, 0)$.

Statement 1 implies statement 2.

However, given that the foci are at $(\pm 1\frac{1}{2}, 0)$ it is not possible to find the eccentricity, e, or the values of a and b, i.e., if $ae = \frac{3}{2}$ then $e = \frac{1}{2}$, $a = 3$ might be another set of values which would give a different equation.

Statement 2 does not imply statement 1. $\hspace{3cm}$ **ANSWER C**

Note that if the directrices are given as well it becomes possible to determine the equation.

The directrices are given by $x = \pm \dfrac{a}{e}$ and so in this case if

$ae = \frac{3}{2}$ and $\dfrac{a}{e} = \frac{25}{6}$ we have $\dfrac{3}{2e} = \dfrac{25e}{6} \Rightarrow e^2 = \frac{9}{25}$.

This gives $e = \frac{3}{5}$ and hence $a = \frac{5}{2}$, $b = 2$.

Example 11

1 The general equation of a hyperbola is $\dfrac{x^2}{a^2} - \dfrac{y^2}{b^2} = 1$.

2 The parametric equations are of the form $x = a \sec \theta$, $y = b \tan \theta$.

The equations $x = a \sec \theta$, $y = b \tan \theta$ can be combined by eliminating θ to give (using $1 + \tan^2\theta = \sec^2\theta$)

$$1 + \left(\frac{y}{b}\right)^2 = \left(\frac{x}{a}\right)^2 \Rightarrow \frac{x^2}{a^2} - \frac{y^2}{b^2} = 1$$

So the parametric form given does imply the standard equation. Statement 2 implies statement 1. However, statement 1 does not necessarily imply statement 2 since there are other parametric forms, e.g., $x = a \cosh \theta$, $y = b \sinh \theta$.   ANSWER B

Short questions

These questions illustrate some of the points that may be raised as a short problem in itself or as part of a longer question.

Example 12

Find the coordinates of the points of intersection of the circles

$$x^2 + y^2 + 2x + 4y - 3 = 0 \quad \text{and} \quad x^2 + y^2 + 3x + 5y = 0$$

If $x^2 + y^2 + 2x + 4y - 3 = 0$ and $x^2 + y^2 + 3x + 5y = 0$
then, subtracting we have

$$-x - y - 3 = 0 \Rightarrow y = -3 - x \tag{1}$$

Substituting in the first equation

$$x^2 + (-3 - x)^2 + 2x + 4(-3 - x) - 3 = 0$$
$$\therefore \qquad 2x^2 + 4x - 6 = 0 \Rightarrow x^2 + 2x - 3 = 0$$
$$\therefore \qquad (x + 3)(x - 1) = 0 \Rightarrow x = -3 \text{ or } x = 1.$$

If $x = -3$, $y = 0$ and if $x = 1$, $y = -4$ (using (1)).
The coordinates of the points of intersection are $(-3, 0)$, $(1, -4)$.

Note that the line $x+y = -3$ is called the radical axis of the two circles. In general, if $S_1 = 0$ and $S_2 = 0$ are the equations of two intersecting circles then $S_1 + \lambda S_2 = 0$ represents a circle through the points of intersection for all $\lambda \neq -1$. When $\lambda = -1$, $S_1 - S_2 = 0$ is a straight line.

Example 13

Find the equation of the circle through the points $A(1,1)$, $B(2,0)$ and $C(-1,-3)$.

Let the equation of the circle be $x^2 + y^2 + 2ax + 2by + c = 0$.

$A(1,1)$ lies on the circle	$2a + 2b + c + 2 = 0$	(1)
$B(2,0)$ lies on the circle	$4a + c + 4 = 0$	(2)
$C(-1,-3)$ lies on the circle	$-2a - 6b + c + 10 = 0$	(3)
Eliminating c from (1) and (2)	$2a - 2b + 2 = 0$	(4)
Eliminating c from (2) and (3)	$-6a - 6b + 6 = 0$	(5)

Multiplying (4) by 3 and subtracting (5) gives $12a = 0 \Rightarrow a = 0$.
Hence $b = 1$ and $c = -4$.
The equation of the circle is $x^2 + y^2 + 2y - 4 = 0$.

Example 14

Show that if the tangents to the curve $y^2 = 4x$ at the points $P(p^2,2p)$ and $Q(q^2,2q)$ contain an angle of $\frac{3}{4}\pi$ then
$$q = \frac{(p-1)}{(p+1)}.$$

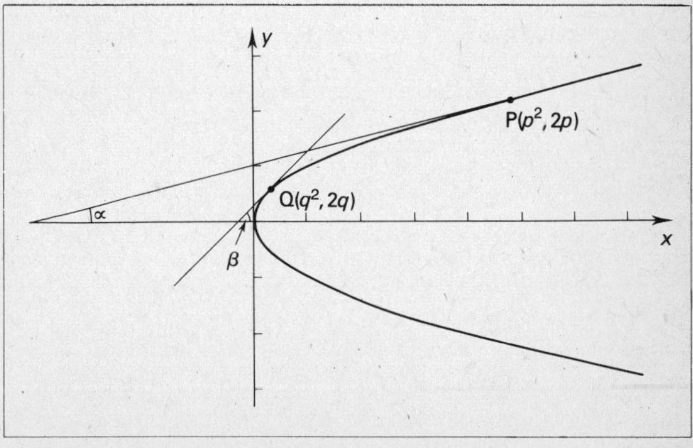

Figure 18

Since $y^2 = 4x$, differentiating implicitly with respect to x

$$2y\frac{dy}{dx} = 4 \Rightarrow \frac{dy}{dx} = \frac{2}{y}$$

Hence the gradients of the tangents at P and Q are $\dfrac{1}{p}$ and $\dfrac{1}{q}$ respectively.

If α and β represent the angles that the tangents make with the x-axis then $\tan\alpha = \dfrac{1}{p}$ and $\tan\beta = \dfrac{1}{q}$.

But $\beta - \alpha = 45°$ $\quad\therefore\quad \tan(\beta - \alpha) = \tan 45°$

$$\therefore \quad \frac{\tan\beta - \tan\alpha}{1 + \tan\beta\tan\alpha} = \frac{\dfrac{1}{q} - \dfrac{1}{p}}{1 + \dfrac{1}{q}\cdot\dfrac{1}{p}} = \frac{p - q}{pq + 1} = 1$$

Simplifying we have

$$pq + 1 = p - q \Rightarrow q(p+1) = p - 1 \quad \text{or} \quad q = \frac{p-1}{p+1}.$$

Example 15

Show that if $P(ap^2, 2ap)$ and $Q(aq^2, 2aq)$ are two points on the parabola $y^2 = 4ax$ such that PQ is a focal chord then the area of the triangle OPQ is $a^2(p-q)$ where O is the vertex of the parabola.

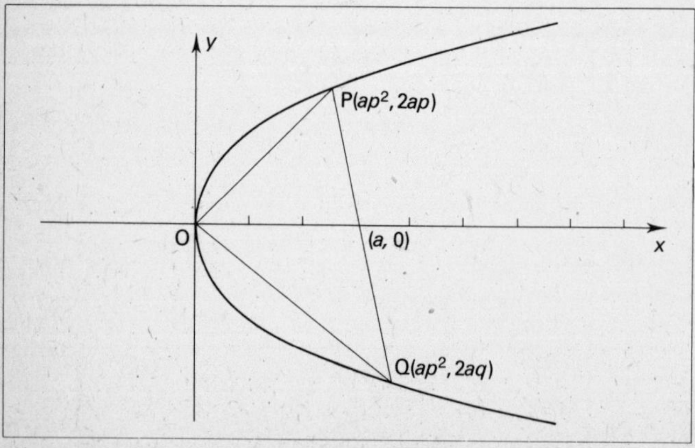

Figure 19

The gradient of the chord $PQ = \dfrac{2ap - 2aq}{ap^2 - aq^2} = \dfrac{2a(p-q)}{a(p^2 - q^2)} = \dfrac{2}{p+q}.$

The equation of the chord PQ is

$$y - 2ap = \frac{2}{p+q}(x - ap^2) \quad \text{or} \quad 2x - (p+q)y + 2apq = 0$$

Since this is a focal chord the point $(a,0)$ lies on the line.

$$\therefore \qquad 2a + 2apq = 0 \Rightarrow pq = -1$$

$\therefore$ the equation of the chord is $2x - (p+q)y - 2a = 0$.

The distance of the origin from this line is given by

$$ON = \frac{2a}{\sqrt{4 + (p+q)^2}}$$

The length of $PQ = \sqrt{\{2a(p-q)\}^2 + (ap^2 - aq^2)^2}$

$$= a(p-q)\sqrt{4 + (p+q)^2}$$

$\therefore$ area of triangle $OPQ = \frac{1}{2}PQ \cdot ON$

$$= \tfrac{1}{2}a(p-q)\sqrt{4 + (p+q)^2} \times \frac{2a}{\sqrt{4 + (p+q)^2}}$$

$$= a^2(p-q)$$

Alternatively it is possible to use the gradients of a focal chord

$$\frac{2ap - 0}{ap^2 - a} = \frac{2aq - 0}{aq^2 - a} \Rightarrow p(q^2 - 1) = q(p^2 - 1)$$

$$\Rightarrow pq = -1$$

Now the area of a triangle whose vertices are $(0,0)$, (x_1, y_1) and (x_2, y_2) is $\frac{1}{2}(x_1 y_2 - x_2 y_1)$.

In this case,

$$\text{area} = \tfrac{1}{2}(2a^2 p^2 q - 2a^2 pq^2) = a^2(p-q) \text{ since } pq = -1$$

Long questions

Many examination questions are of a fairly lengthy nature and may involve several aspects of, say, one of the conics. The derivation of the standard equations of tangents, normals, etc. often forms the first part of such questions.

> **Example 16**
>
> Find the equation of the normal to the ellipse $\dfrac{x^2}{a^2}+\dfrac{y^2}{b^2}=1$ at the point $P(a\cos\theta, b\sin\theta)$. If the normal cuts the axes at A and B find the coordinates of A and B and deduce that, as P varies, the locus of the mid-point of AB is also an ellipse.

Since $\dfrac{x^2}{a^2}+\dfrac{y^2}{b^2}=1$, differentiating with respect to x

$$\frac{2x}{a^2}+\frac{2y}{b^2}\frac{dy}{dx}=0 \Rightarrow \frac{dy}{dx}=-\frac{b^2x}{a^2y}$$

The gradient of the tangent at $P(a\cos\theta, b\sin\theta)$ is $-\dfrac{b\cos\theta}{a\sin\theta}$.

Hence the gradient of the normal at P is $\dfrac{a\sin\theta}{b\cos\theta}$.

The equation of the normal at P is

$$y-b\sin\theta=\frac{a\sin\theta}{b\cos\theta}(x-a\cos\theta)$$

or

$$ax\sin\theta-by\cos\theta=(a^2-b^2)\sin\theta\cos\theta$$

If this normal cuts the x-axis at A and the y-axis at B then,

A is the point $\left(\dfrac{a^2-b^2}{a}\cos\theta,0\right)$ B is the point $\left(0,-\dfrac{a^2-b^2}{b}\sin\theta\right)$.

Let $M(x,y)$ be the mid-point of AB.

Hence $x=\dfrac{(a^2-b^2)}{2a}\cos\theta$ and $y=-\dfrac{(a^2-b^2)}{2b}\sin\theta$.

$$\therefore \cos\theta=\frac{2ax}{(a^2-b^2)} \quad \text{and} \quad \sin\theta=-\frac{2by}{(a^2-b^2)}$$

Using the identity $\cos^2\theta+\sin^2\theta=1$ we have

$$\frac{4a^2x^2}{(a^2-b^2)^2}+\frac{4b^2y^2}{(a^2-b^2)^2}=1$$

This is of the form $\dfrac{x^2}{c^2}+\dfrac{y^2}{d^2}=1$ and hence the locus of the mid-point of AB is an ellipse.

Example 17

Show that the equation of the normal to the rectangular hyperbola $xy = c^2$ at the point $P\left(ct, \dfrac{c}{t}\right)$ is

$$t^3x - ty + c(1 - t^4) = 0.$$

Find the coordinates of the point Q where this normal meets the curve again. Hence find the locus of the mid-point of PQ as P varies.

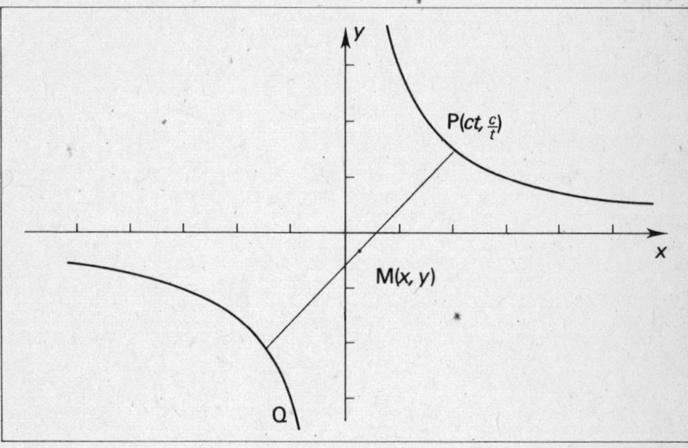

Figure 20

Differentiating $xy = c^2$ implicitly with respect to x

$$x\frac{dy}{dx} + y = 0 \Rightarrow \frac{dy}{dx} = -\frac{y}{x}$$

Thus the gradient of the tangent at $P\left(ct, \dfrac{c}{t}\right)$ is $-\dfrac{1}{t^2}$.

The gradient of the normal at P is t^2.

The equation of the normal at P is

$$y - \frac{c}{t} = t^2(x - ct) \quad \text{or} \quad t^3x - ty + c(1 - t^4) = 0$$

This line cuts the curve $xy = c^2$ again at Q and hence, solving the

91

two equations simultaneously

$$t^3x - \frac{tc^2}{x} + c(1-t^4) = 0 \Rightarrow t^3x^2 + c(1-t^4)x - c^2t = 0$$

Since $x = ct$ is a solution $(x-ct)(t^3x+c) = 0$.

Hence $x = -\dfrac{c}{t^3}$ is the x coordinate of $Q \Rightarrow Q$ is at $\left(-\dfrac{c}{t^3}, -ct^3\right)$.

Let $M(x,y)$ be the mid-point of PQ.

Thus $x = \frac{1}{2}\left\{ct + \left(-\dfrac{c}{t^3}\right)\right\}$ and $y = \frac{1}{2}\left\{\dfrac{c}{t} + (ct^3)\right\}$

or $2xt^3 = c(t^4-1)$ (1) and $2yt = c(1-t^4)$ (2)

Combining (1) and (2) $2xt^3 = -2yt$ $t^2 = -\dfrac{y}{x}$

Substituting in (2) after squaring

$$4y^2t^2 = c^2(1-t^4)^2 \Rightarrow 4y^2\left(-\dfrac{y}{x}\right) = c^2\left(1-\dfrac{y^2}{x^2}\right)^2$$

The locus is $4x^3y^3 = c^2(x^2-y^2)$.

It is a useful technique to use the slopes to find the coordinates of Q.

P is the point $\left(ct, \dfrac{c}{t}\right)$ and Q is the point $\left(ct_1, \dfrac{c}{t_1}\right)$.

Slope of chord $PQ = -\dfrac{1}{tt_1}$

Slope of normal at P is t^2.

$$\therefore \quad -\frac{1}{tt_1} = t^2 \Rightarrow t_1 = \frac{-1}{t^3}$$

and the coordinates of Q are $\left(\dfrac{-c}{t^3}, -ct^3\right)$.

Chapter 7

Matrices and Determinants

Multiple choice questions (Type A) Select the correct answer

Example 1
Which matrix represents a rotation?

A $\begin{pmatrix} -1 & 0 \\ 0 & -1 \end{pmatrix}$ **B** $\begin{pmatrix} 0 & -1 \\ -1 & 0 \end{pmatrix}$ **C** $\begin{pmatrix} -1 & 0 \\ 0 & 1 \end{pmatrix}$

D $\begin{pmatrix} 1 & 0 \\ 0 & -1 \end{pmatrix}$ **E** $\begin{pmatrix} 0 & 1 \\ 1 & 0 \end{pmatrix}$

The simplest way to answer the question is to find the matrix whose determinant is equal to 1 as this is a necessary condition for a rotation. The determinant gives the area scale factor. **A** is the only matrix whose determinant is equal to 1. The others all have determinants equal to -1 and represent reflexions.   ANSWER A

In other questions you may have to find the image of the unit square. Remember that the first matrix column gives the image of the point (1,0) and the second column gives the image of (0,1). For matrix **A**, $(1,0) \rightarrow (-1,0)$ and $(0,1) \rightarrow (0,-1)$. This is a rotation of $180°$ about the origin.

Example 2
The inverse of the matrix $\begin{pmatrix} 3 & 1 \\ 4 & 2 \end{pmatrix}$ is

A $\begin{pmatrix} 3 & -1 \\ -4 & 2 \end{pmatrix}$ **B** $\begin{pmatrix} -3 & 1 \\ 4 & -2 \end{pmatrix}$ **C** $\begin{pmatrix} 2 & -1 \\ -4 & 3 \end{pmatrix}$

D $\begin{pmatrix} -2 & 1 \\ 4 & -3 \end{pmatrix}$ **E** $\begin{pmatrix} 1 & -\frac{1}{2} \\ -2 & 1\frac{1}{2} \end{pmatrix}$

The inverse of

$$\begin{pmatrix} a & b \\ c & d \end{pmatrix} \text{ is } \frac{1}{\Delta}\begin{pmatrix} d & -b \\ -c & a \end{pmatrix} = \begin{pmatrix} d/\Delta & -b/\Delta \\ -c/\Delta & a/\Delta \end{pmatrix} \text{ where } \Delta = ad - bc.$$

The inverse of $\begin{pmatrix} 3 & 1 \\ 4 & 2 \end{pmatrix}$ is $\frac{1}{2}\begin{pmatrix} 2 & -1 \\ -4 & 3 \end{pmatrix} = \begin{pmatrix} 1 & -\frac{1}{2} \\ -2 & 1\frac{1}{2} \end{pmatrix}$

ANSWER E

Example 3
Which matrix represents a translation?

A $\begin{pmatrix} 2 & 0 \\ 3 & 0 \end{pmatrix}$ **B** $\begin{pmatrix} 2 & 3 \\ 0 & 0 \end{pmatrix}$ **C** $\begin{pmatrix} 2 & 0 \\ 0 & 3 \end{pmatrix}$ **D** $\begin{pmatrix} 0 & 2 \\ 3 & 0 \end{pmatrix}$

E none of these

For a transformation represented by the matrix $\begin{pmatrix} a & b \\ c & d \end{pmatrix}$, (x,y)

moves to (X,Y) where $\begin{pmatrix} X \\ Y \end{pmatrix} = \begin{pmatrix} a & b \\ c & d \end{pmatrix}\begin{pmatrix} x \\ y \end{pmatrix}$. So $X = ax + by$ and $Y = cx + dy$.

The translation $\begin{pmatrix} 2 \\ 3 \end{pmatrix}$ or $2\mathbf{i} + 3\mathbf{j}$, which takes (x,y) to $(x+2, y+3)$, cannot be represented by premultiplying by a 2×2 matrix. Another reason is that any 2×2 matrix transformation keeps the origin fixed, whereas it would need to move in a translation. ANSWER E

Example 4
Which represents, in three dimensions, a rotation of 180° about the x-axis.

A $\begin{pmatrix} 1 & 0 & 0 \\ 0 & -1 & 0 \\ 0 & 0 & 1 \end{pmatrix}$ **B** $\begin{pmatrix} -1 & 0 & 0 \\ 0 & 1 & 0 \\ 0 & 0 & -1 \end{pmatrix}$ **C** $\begin{pmatrix} 1 & 0 & 0 \\ 0 & -1 & 0 \\ 0 & 0 & -1 \end{pmatrix}$

D $\begin{pmatrix} -1 & 0 & 0 \\ 0 & 1 & 0 \\ 0 & 0 & 1 \end{pmatrix}$ **E** $\begin{pmatrix} -1 & 0 & 0 \\ 0 & -1 & 0 \\ 0 & 0 & 1 \end{pmatrix}$

The three columns of the matrix represent the images of the three points $I(1,0,0)$, $J(0,1,0)$ and $K(0,0,1)$. In a half-turn about the x-axis, I is invariant (which rules out answers B, D and E), J goes to $(0,-1,0)$ and K goes to $(0,0,-1)$, which rules out answer A.   ANSWER C

Example 5

If matrix X has order $p \times q$ and matrix Y has order $q \times r$ then matrix XY has order

A $p \times q$ **B** $q \times r$ **C** $r \times p$ **D** $p \times r$ **E** $q \times p$

Matrix X $\times$ Y $=$ XY

Order $p \times q$ $q \times r$ $p \times r$

In order for X and Y to be multiplied together (to be compatible) the number of columns in X (q) must equal the number of rows in Y (q). The product XY will have p rows and r columns, i.e., order $p \times r$.   ANSWER D

Example 6

If matrix X has order $p \times q$, matrix Y has order $q \times r$ and matrix Z has order $r \times s$ then matrix XYZ has order

A $p \times q$ **B** $q \times r$ **C** $r \times s$ **D** $p \times r$ **E** $p \times s$

Matrix multiplication is associative so $(XY)Z = X(YZ)$. From Example 5, the order of XY is $p \times r$, so the order of $(XY)Z$ is $p \times s$. Similarly, the order of YZ is $q \times s$, so the order of $X(YZ)$ is $p \times s$.   ANSWER E

Example 7

Which matrix does not represent an enlargement?

A $\begin{pmatrix} 3 & 0 \\ 0 & 3 \end{pmatrix}$ **B** $\begin{pmatrix} \frac{1}{2} & 0 \\ 0 & \frac{1}{2} \end{pmatrix}$ **C** $\begin{pmatrix} -2 & 0 \\ 0 & -2 \end{pmatrix}$ **D** $\begin{pmatrix} 3 & 0 & 0 \\ 0 & 3 & 0 \\ 0 & 0 & 3 \end{pmatrix}$

E $\begin{pmatrix} -2 & 0 & 0 \\ 0 & -2 & 0 \\ 0 & 0 & -2 \end{pmatrix}$

Matrix A takes (x,y) to $(3x,3y)$ and multiplies lengths by 3 and areas by 9. A mathematical 'enlargement' can also reduce a figure in size, e.g., matrix B will reduce dimensions by one half and the area by one quarter of the original size. Matrix C is a negative enlargement which in two dimensions is equivalent to a rotation of $180°$ followed by an enlargement of scale factor $+2$. Matrix D is a three-dimensional enlargement where lengths are tripled, areas multiplied by 9 and volumes by 27. Matrix E, in addition to doubling lengths, quadrupling areas and multiplying volumes by 8, also turns a three-dimensional figure inside out and destroys the sense of the transformation. Matrix E is therefore not an enlargement.   ANSWER E

Multiple choice questions (Type B) Answer according to the table

A	B	C	D	E
1, 2, 3	1, 3	2, 3	2	3
correct	only	only	only	only

Example 8

The determinant of the matrix $\begin{pmatrix} a & b \\ c & d \end{pmatrix}$ is

1 $ab-cd$ **2** $ac-bd$ **3** $ad-bc$

This is simply definition: statement 3 only is correct.

ANSWER E

Example 9

The determinant of the matrix $\begin{pmatrix} a & b & c \\ d & e & f \\ g & h & k \end{pmatrix}$ is equal to

1 $a(ek-fh)+b(dk-fg)+c(dh-ge)$
2 $a(ek-fh)+b(fg-dk)+c(dh-ge)$
3 $a(ek-fh)-b(dk-fg)+c(dh-ge)$

Statement 3 is the form usually remembered and statement 2 is the same as this but slightly rearranged. Statements 2 and 3 are correct.

ANSWER C

Example 10

$$\begin{pmatrix} 1 & 2 \\ 2 & 4 \end{pmatrix}\begin{pmatrix} a & b \\ c & d \end{pmatrix} = \begin{pmatrix} 1 & 0 \\ 0 & 1 \end{pmatrix}$$

$$\quad A \qquad\quad B \qquad\qquad C$$

1 C is the identity matrix.   **2** B is the inverse of A.
3 The values of a, b, c and d cannot be found.

Statement 1 is correct:

$$\begin{pmatrix} 1 & 0 \\ 0 & 1 \end{pmatrix}\begin{pmatrix} a & b \\ c & d \end{pmatrix} = \begin{pmatrix} a & b \\ c & d \end{pmatrix} = \begin{pmatrix} a & b \\ c & d \end{pmatrix}\begin{pmatrix} 1 & 0 \\ 0 & 1 \end{pmatrix}$$

If the values of a, b, c and d could be found then B would be the inverse of A, but these values cannot be found so A has no inverse. $a + 2c = 1$ and $2a + 4c = 0$, which are inconsistent.
Statements 1 and 3 are correct.   ANSWER B
In general, a matrix cannot have an inverse if its determinant is zero.

Example 11
If $AB = I$ where A and B are matrices and I is the 2×2 identity matrix

1 A and B have the same order.   **2** B is the inverse of A.
3 A and B are compatible.

Statement 3 means that A and B can be multiplied together (which is true) and this can happen if A is $\begin{pmatrix} 1 & 2 & 0 \\ 2 & 5 & 0 \end{pmatrix}$ and B is $\begin{pmatrix} 5 & -2 \\ -2 & 1 \\ 3 & 4 \end{pmatrix}$, i.e., A and B are of different order.
Only statement 3 is correct.   ANSWER E

Multiple choice questions (Type C) For each question two statements are given. Answer

A if 1 implies 2 and 2 implies 1
B if 2 implies 1 but 1 does not imply 2
C if 1 implies 2 but 2 does not imply 1
D if 1 denies 2 and 2 denies 1
E if none of these relations holds

Example 12

For the matrix $\begin{pmatrix} a & b \\ c & d \end{pmatrix}$

1 $ad = bc$ **2** M has an inverse matrix

The inverse of M, M^{-1}, satisfies $MM^{-1} = M^{-1}M = I = \begin{pmatrix} 1 & 0 \\ 0 & 1 \end{pmatrix}$.

$M^{-1} = \dfrac{1}{\Delta}\begin{pmatrix} d & -b \\ -c & a \end{pmatrix}$ and if $\Delta = 0$ no matrix M^{-1} can be found to satisfy $MM^{-1} = I$. $ad = bc \Rightarrow \Delta = ad - bc = 0$.

Statement 1 denies statement 2 and vice versa. **ANSWER D**

Example 13

A is a non-singular matrix

1 B is the inverse of A **2** $AB = BA$

A singular matrix has determinant zero and so has no inverse. If B is the inverse of A then $AB = BA = I$ (by definition) so statement 1 implies statement 2.

$AB = BA$ is not true in general for matrices (matrix multiplication is not commutative) but the result is true for some particular matrices, e.g., $AI = IA = A$; $AA^{-1} = A^{-1}A = I$.

(Remember I is a square matrix and A and I have the same order.)

$$\begin{pmatrix} 3 & -4 \\ 4 & 3 \end{pmatrix}\begin{pmatrix} 5 & -12 \\ 12 & 5 \end{pmatrix} = \begin{pmatrix} -33 & -56 \\ 56 & -33 \end{pmatrix} = \begin{pmatrix} 5 & -12 \\ 12 & 5 \end{pmatrix}\begin{pmatrix} 3 & -4 \\ 4 & 3 \end{pmatrix}$$

Both of these matrices represent a rotation combined with an enlargement and the order of application does not affect the result.

$$\begin{pmatrix} 1 & 2 \\ 3 & 4 \end{pmatrix}\begin{pmatrix} 3 & 2 \\ 3 & 6 \end{pmatrix} = \begin{pmatrix} 9 & 14 \\ 21 & 30 \end{pmatrix} = \begin{pmatrix} 3 & 2 \\ 3 & 6 \end{pmatrix}\begin{pmatrix} 1 & 2 \\ 3 & 4 \end{pmatrix}$$

These matrices seem general enough, but if the first is A then the second is $A + 2I$ and

$$A(A+2I) = A^2 + 2AI = A^2 + 2A$$
$$(A+2I)A = A^2 + 2IA = A^2 + 2A$$

$AB = BA$ does not necessarily imply that B is the inverse of A but it could be. Statement 2 does not imply statement 1. **ANSWER C**

> **Example 14**
> A and B are square matrices
> **1** $\det A \times \det B = 1$ **2** $AB = I$

$AB = I \Rightarrow \det(AB) = \det I = 1$
Since $\det(AB) = \det A \times \det B$, statement 1 is therefore true so statement 2 implies statement 1.
$\det A \times \det B = \det(AB) = 1$ but there are many matrices whose determinants are 1 and AB could be any of these.
Statement 1 does not imply statement 2.   **ANSWER B**

> **Example 15**
> A and B are square matrices
> **1** AB is the zero matrix   **2** Either A or B is the zero matrix

If either A or B is zero so AB will be zero. Therefore statement 2 implies statement 1.
AB could be zero without A or B being zero, e.g.,

$$\begin{pmatrix} 1 & 0 \\ 2 & 0 \end{pmatrix}\begin{pmatrix} 0 & 0 \\ 2 & 1 \end{pmatrix} = \begin{pmatrix} 0 & 0 \\ 0 & 0 \end{pmatrix}$$

Statement 1 does not imply statement 2.   **ANSWER B**

> **Example 16**
> **1** $(AB)^{-1} = A^{-1}B^{-1}$ **2** $(AB)^{-1} = B^{-1}A^{-1}$

Provided A and B have inverses (without which the question is meaningless) statement 2 is always true, since

$$(AB)^{-1}AB = B^{-1}A^{-1}AB = B^{-1}IB = B^{-1}B = I$$

If $(AB)^{-1} = A^{-1}B^{-1}$ then $(AB)^{-1}AB = I \Rightarrow A^{-1}B^{-1}AB = I$
Premultiplying by A gives $AA^{-1}B^{-1}AB = AI \Rightarrow B^{-1}AB = A$
Premultiplying by B gives $BB^{-1}AB = BA \Rightarrow AB = BA$

This implies A and B are commutative which is true only for a certain type of matrix (see Example 13) and is not true in general.
So statement 1 implies statement 2 but statement 2 does not necessarily imply statement 1.   **ANSWER C**

> **Example 17**
> If A and B are 2×2 matrices
> **1** A and B represent rotations   **2** AB represents a rotation

If A is a 2×2 matrix representing a rotation then the centre is the origin and the product of two rotations about the origin will be a third whose angle of rotation will be the sum of the first two rotations.

However the product of two reflexions in lines through the origin will be a rotation so statement 2 is true if A and B are reflexions. Statement 1 implies statement 2 but statement 2 does not imply statement 1. ANSWER C

Multiple choice questions (Type D) Each question consists of a problem followed by four pieces of information. Decide whether the problem can be solved with one of the four pieces of information omitted and answer

A if 1 could be omitted   **D** if 4 could be omitted
B if 2 could be omitted   **E** if none can be omitted
C if 3 could be omitted

Example 18

Find the value of the determinant of the matrix $\begin{pmatrix} a & b \\ c & d \end{pmatrix}$

1 $a = 1$ **2** $b = 2$ **3** $c = 3$ **4** $d = 4$

The determinant of the matrix is $ad - bc$ so all four values are needed. ANSWER E

Example 19

Find the value of the determinants of the matrix $\begin{pmatrix} a & 0 & d \\ 0 & b & 0 \\ 0 & 1 & c \end{pmatrix}$

1 $a = 1$ **2** $b = 2$ **3** $c = 3$ **4** $d = 4$

The determinant is equal to $a(bc - 0) + d(0 - 0) = abc$. Value 4 $(d = 4)$ can be omitted. ANSWER D

Example 20

The matrix $\begin{pmatrix} a & b \\ c & d \end{pmatrix}$ represents a reflexion in a line through the origin.

1 $a = \frac{3}{5}$ **2** $a + d = 0$ **3** $b = c$ **4** $a^2 + b^2 = 1$

The matrix $\begin{pmatrix} \cos 2\theta & \sin 2\theta \\ \sin 2\theta & -\cos 2\theta \end{pmatrix}$ represents a reflexion in the line $y = x \tan \theta$ (see Example 22).

Values 2 and 3 ensure that the matrix is of the form $\begin{pmatrix} a & b \\ b & -a \end{pmatrix}$ and to ensure that a can be a cosine and b the sine of the same angle we need $a^2 + b^2 = 1$. The fact that $a = \frac{3}{5}$ is not needed to make sure it is a rotation.   ANSWER A

$A = \frac{3}{5}$ specifies where the line of reflexion is, or rather where two possible lines of reflexion are because $a = \frac{3}{5} \Rightarrow b = \pm\frac{4}{5}$.

Short questions

> ### Example 21
> Show that for a transformation in the xy plane represented by the matrix $\begin{pmatrix} a & b \\ c & d \end{pmatrix}$ the area is multiplied by $\Delta = ad - bc$, the determinant of the matrix.

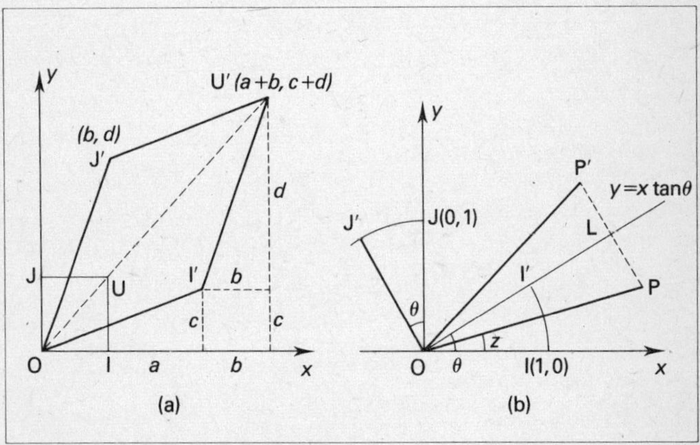

Figure 21

In Figure 21(a) the unit square $OIUJ$ (area 1) is transformed into the parallelogram $OI'U'J'$. The coordinates can be evaluated by matrix multiplication and $\mathbf{OJ'} = b\mathbf{i} + d\mathbf{j} = \mathbf{I'U'}$.

Area of $OI'U'J' = 2 \times$ area triangle $OI'U'$

$$= 2 \times \{\tfrac{1}{2}(a+b)(c+d) - \tfrac{1}{2}ac - bc - \tfrac{1}{2}bd\}$$

$$= ac + ad + bc + bd - ac - 2bc - bd = ad - bc = \Delta$$

The area scale factor is $\Delta = ad - bc$.

Example 22

Find the matrix R which represents a rotation of $+\theta°$ about the origin, and the matrix M, which represents a reflexion in the line $y = x \tan \theta$ (both in two dimensions).

With the notation in Figure 21(b) the point $I(1,0)$ rotates through θ to the point $I'(\cos \theta, \sin \theta)$ and $J(0,1)$ rotates to $J'(-\sin \theta, \cos \theta)$.

$$R = \begin{pmatrix} \cos \theta & -\sin \theta \\ \sin \theta & \cos \theta \end{pmatrix} \qquad \det R = \cos^2\theta + \sin^2\theta = 1$$

$$\text{area is preserved (invariant)}$$

For M, consider the image P' of the point $P(r \cos z, r \sin z)$.
The length of $OP = r \Rightarrow OP' = r$ (reflexion).
$\angle POL = \theta - z = \angle LOP'$ so $\angle IOP' = \theta + \theta - z = 2\theta - z$.
P' is $[r \cos(2\theta - z), r \sin(2\theta - z)]$
M takes $P(r \cos z, r \sin z)$ to $P'[r \cos(2\theta - z), r \sin(2\theta - z)]$

$$M\begin{pmatrix} r \cos z \\ r \sin z \end{pmatrix} = \begin{pmatrix} r \cos(2\theta - z) \\ r \sin(2\theta - z) \end{pmatrix} = \begin{pmatrix} r \cos 2\theta \cos z + r \sin 2\theta \sin z \\ r \sin 2\theta \cos z - r \cos 2\theta \sin z \end{pmatrix}$$

$$= \begin{pmatrix} \cos 2\theta & \sin 2\theta \\ \sin 2\theta & -\cos 2\theta \end{pmatrix}\begin{pmatrix} r \cos z \\ r \sin z \end{pmatrix}$$

$$M = \begin{pmatrix} \cos 2\theta & \sin 2\theta \\ \sin 2\theta & -\cos 2\theta \end{pmatrix} \qquad \det M = -\cos^2 2\theta - \sin^2 2\theta = -1$$

The area is invariant, the negative sign indicating a reflexion (rather than a rotation).

Example 23

Demonstrate the result that $\det(AB) = \det A \times \det B$ by finding the determinants of the 2×2 matrices $A = \begin{pmatrix} a & b \\ c & d \end{pmatrix}$

$B = \begin{pmatrix} e & f \\ g & h \end{pmatrix}$ and AB.

$$AB = \begin{pmatrix} a & b \\ c & d \end{pmatrix}\begin{pmatrix} e & f \\ g & h \end{pmatrix} = \begin{pmatrix} ae + bg & af + bh \\ ce + dg & cf + dh \end{pmatrix}$$

$$\det(AB) = (ae+bg)(cf+dh) - (af+bh)(ce+dg)$$

$$= aecf + aedh + bgcf + bgdh - afce - afdg - bhce - bhdg$$

$$= aedh - adfg - bceh + bgcf = ad(eh-fg) - bc(eh-fg)$$

$$= (ad-bc)(eh-fg)$$

$$= \det A \times \det B$$

Example 24

Given
$$A = \begin{pmatrix} 1 & 2 & 3 \\ 2 & 0 & 1 \end{pmatrix} \quad B = \begin{pmatrix} 2 & 3 \\ -1 & 2 \\ 2 & 1 \end{pmatrix} \quad C = \begin{pmatrix} 1 \\ 3 \end{pmatrix} \text{ find}$$

$(AB)C$ and $A(BC)$ and comment.

$$(AB)C = \left[\begin{pmatrix} 1 & 2 & 3 \\ 2 & 0 & 1 \end{pmatrix} \begin{pmatrix} 2 & 3 \\ -1 & 2 \\ 2 & 1 \end{pmatrix} \right] \begin{pmatrix} 1 \\ 3 \end{pmatrix} = \begin{pmatrix} 6 & 10 \\ 6 & 7 \end{pmatrix} \begin{pmatrix} 1 \\ 3 \end{pmatrix} = \begin{pmatrix} 36 \\ 27 \end{pmatrix}$$

$$A(BC) = \begin{pmatrix} 1 & 2 & 3 \\ 2 & 0 & 1 \end{pmatrix} \left[\begin{pmatrix} 2 & 3 \\ -1 & 2 \\ 2 & 1 \end{pmatrix} \begin{pmatrix} 1 \\ 3 \end{pmatrix} \right] = \begin{pmatrix} 1 & 2 & 3 \\ 2 & 0 & 1 \end{pmatrix} \begin{pmatrix} 11 \\ 5 \\ 5 \end{pmatrix}$$

$$= \begin{pmatrix} 36 \\ 27 \end{pmatrix}$$

$A(BC) = (AB)C$

This is one example which shows that matrix multiplication is associative.

Example 25

Use the matrix for a rotation of θ to transform the point $(\cos z, \sin z)$ to the point $(\cos(z+\theta), \sin(z+\theta))$ and deduce the addition formulae.

The point $P(\cos z, \sin z)$ is distant 1 from the origin and when rotated through an angle θ to Q, OQ makes an angle $(z+\theta)$ with the x-axis. Since OQ has length 1, the coordinates of Q are

$$(\cos(z+\theta), \sin(z+\theta))$$

and

$$\begin{pmatrix} \cos(z+\theta) \\ \sin(z+\theta) \end{pmatrix} = \begin{pmatrix} \cos\theta & -\sin\theta \\ \sin\theta & \cos\theta \end{pmatrix}\begin{pmatrix} \cos z \\ \sin z \end{pmatrix}$$

$$= \begin{pmatrix} \cos z \cos\theta - \sin z \sin\theta \\ \sin z \cos\theta + \cos z \sin\theta \end{pmatrix}$$

Example 26
Prove the result for matrices $(AB)^{-1} = B^{-1}A^{-1}$

Knowing that matrix multiplication is associative

$$B^{-1}A^{-1}AB = B^{-1}IB \qquad \text{since } A^{-1}A = I$$
$$= B^{-1}B = I$$

$(AB)^{-1}$ is the inverse of AB so
$$(AB)^{-1}AB = I \Rightarrow (AB)^{-1} = B^{-1}A^{-1}.$$

Alternatively $(AB)^{-1}AB = I \Rightarrow (AB)^{-1}ABB^{-1} = IB^{-1}$
$$\Rightarrow (AB)^{-1}A = B^{-1}$$
$$\Rightarrow (AB)^{-1}AA^{-1} = B^{-1}A^{-1}$$
$$\Rightarrow (AB)^{-1} = B^{-1}A^{-1}$$

Long questions

Example 27
Find the inverse of $\begin{pmatrix} 1 & 2 & 3 \\ 2 & 5 & -1 \\ 3 & 8 & -4 \end{pmatrix}$ and hence or otherwise

solve
$$\begin{aligned} x+2y+3z &= 14 \\ 2x+5y-z &= 9 \\ 3x+8y-4z &= 7 \end{aligned}$$

Method 1 The inverse of

$$M = \begin{pmatrix} a & b & c \\ d & e & f \\ g & h & i \end{pmatrix} \text{ is } \frac{1}{\Delta}\begin{pmatrix} A & D & G \\ B & E & H \\ C & F & I \end{pmatrix} \text{ where } \begin{pmatrix} A & B & C \\ D & E & F \\ G & H & I \end{pmatrix}$$

is the matrix of the cofactors of M. Each element of M has a cofactor (denoted by the corresponding capital letter) which is the value of the 2×2 determinant left by omitting the row and column containing the letter (remembering to change the sign of B, D, F and H). $\Delta = \det M$, e.g.,

$$A = \begin{vmatrix} e & f \\ h & i \end{vmatrix} = ei - fh \qquad B = -\begin{vmatrix} d & f \\ g & i \end{vmatrix} = -(di - fg)$$

$A = -20 + 8 = -12; \quad B = -(-8 + 3) = 5; \quad C = 16 - 15 = 1;$
$D = -(-8 - 24) = 32; \quad E = -4 - 9 = -13; \quad F = -(8 - 6) = -2;$
$G = -2 - 15 = -17; \quad H = -(-1 - 6) = 7; \quad I = 5 - 4 = 1$
$\Delta = \det M = 1(-20 + 8) - 2(-8 + 3) + 3(16 - 15) = -12 + 10 + 3 = 1$

$$M^{-1} = \begin{pmatrix} -12 & 32 & -17 \\ 5 & -13 & 7 \\ 1 & -2 & 1 \end{pmatrix} \quad \text{and} \quad \begin{pmatrix} x \\ y \\ z \end{pmatrix} = M^{-1} \begin{pmatrix} 14 \\ 9 \\ 7 \end{pmatrix}$$

$$= \begin{pmatrix} -12 & 32 & -17 \\ 5 & -13 & 7 \\ 1 & -2 & 1 \end{pmatrix} \begin{pmatrix} 14 \\ 9 \\ 7 \end{pmatrix} = \begin{pmatrix} 1 \\ 2 \\ 3 \end{pmatrix}$$

Method 2 Solve the equations
$$\begin{aligned} x + 2y + 3z &= 14 &\quad (1) \\ 2x + 5y - z &= 9 &\quad (2) \\ 3x + 8y - 4z &= 7 &\quad (3) \end{aligned}$$

$(2) - 2(1)$ gives $y - 7z = -19$ (4)
$(3) - 3(1)$ gives $2y - 13z = -35$ (5)

$(5) - 2(4)$ gives $z = 3$

Substituting back in (4) gives $y = 7z - 19 = 21 - 19 = 2 \Rightarrow x = 1.$

Many students go round in circles with this method because they forget to stick to a pattern:

1. Eliminate x from (1) and (2).
2. Eliminate x from (1) and (3).
3. Eliminate y from (4) and (5) to give z, then find y and x.

Method 3 Do the same row operations (as in method 2) to reduce the coefficient matrix to echelon form and then to the identity. If the same operations are done to the identity the inverse is obtained.

$$\begin{pmatrix} 1 & 2 & 3 \\ 2 & 5 & -1 \\ 3 & 8 & -4 \end{pmatrix} \begin{pmatrix} 14 \\ 9 \\ 7 \end{pmatrix} \begin{pmatrix} 1 & 0 & 0 \\ 0 & 1 & 0 \\ 0 & 0 & 1 \end{pmatrix}$$

Row $2 - 2 \times$ Row 1
Row $3 - 3 \times$ Row 1

$$\begin{pmatrix} 1 & 2 & 3 \\ 0 & 1 & -7 \\ 0 & 2 & -13 \end{pmatrix} \begin{pmatrix} 14 \\ -19 \\ -35 \end{pmatrix} \begin{pmatrix} 1 & 0 & 0 \\ -2 & 1 & 0 \\ -3 & 0 & 1 \end{pmatrix}$$

Row $3 - 2 \times$ Row 2

$$\begin{pmatrix} 1 & 2 & 3 \\ 0 & -1 & -7 \\ 0 & 0 & 1 \end{pmatrix} \begin{pmatrix} 14 \\ -19 \\ 3 \end{pmatrix} \begin{pmatrix} 1 & 0 & 0 \\ -2 & 1 & 0 \\ 1 & -2 & 1 \end{pmatrix}$$

Row $2 + 7 \times$ Row 3
$$\begin{pmatrix} 1 & 2 & 3 \\ 0 & 1 & 0 \\ 0 & 0 & 1 \end{pmatrix} \begin{pmatrix} 14 \\ 2 \\ 3 \end{pmatrix} \begin{pmatrix} 1 & 0 & 0 \\ 5 & -13 & 7 \\ 1 & -2 & 1 \end{pmatrix}$$

Row $1 - 2 \times$ Row $2 - 3 \times$ Row 3
$$\begin{pmatrix} 1 & 0 & 0 \\ 0 & 1 & 0 \\ 0 & 0 & 1 \end{pmatrix} \begin{pmatrix} 1 \\ 2 \\ 3 \end{pmatrix} \begin{pmatrix} -12 & 32 & -17 \\ 5 & -13 & 7 \\ 1 & -2 & 1 \end{pmatrix}$$

Identity   Solution   Inverse

Example 28

Find the solution of the three equations $x + 2y + 3z = 4$, $2x + 3y + 4z = 2$ and $3x + 5y + pz = r$ in the cases where
(a) $p = 8$ and $r = 8$ (b) $p = 7$ and $r = 8$ (c) $p = 7$ and $r = 6$.

(a)

$$\begin{pmatrix} 1 & 2 & 3 \\ 2 & 3 & 4 \\ 3 & 5 & 8 \end{pmatrix} \begin{pmatrix} x \\ y \\ z \end{pmatrix} = \begin{pmatrix} 4 \\ 2 \\ 8 \end{pmatrix} \Rightarrow \begin{pmatrix} 1 & 2 & 3 \\ 0 & -1 & -2 \\ 0 & -1 & -1 \end{pmatrix} \begin{pmatrix} x \\ y \\ z \end{pmatrix} = \begin{pmatrix} 4 \\ -6 \\ -4 \end{pmatrix}$$

$$\Rightarrow \begin{pmatrix} 1 & 2 & 3 \\ 0 & -1 & -2 \\ 0 & 0 & 1 \end{pmatrix} \begin{pmatrix} x \\ y \\ z \end{pmatrix} = \begin{pmatrix} 4 \\ -6 \\ 2 \end{pmatrix}$$

$\Rightarrow$ third equation is $1 \times z = 2 \Rightarrow z = 2$
second equation is $y + 2z = 6 \Rightarrow y = 2$
first equation is $x + 2y + 3z = 4 \Rightarrow x = -6$

The three equations represent three planes which intersect at the point $(-6, 2, 2)$.

(b)

$$\begin{pmatrix} 1 & 2 & 3 \\ 2 & 3 & 4 \\ 3 & 5 & 7 \end{pmatrix} \begin{pmatrix} x \\ y \\ z \end{pmatrix} = \begin{pmatrix} 4 \\ 2 \\ 8 \end{pmatrix} \Rightarrow \begin{pmatrix} 1 & 2 & 3 \\ 0 & -1 & -2 \\ 0 & -1 & -2 \end{pmatrix} \begin{pmatrix} x \\ y \\ z \end{pmatrix} = \begin{pmatrix} 4 \\ -6 \\ -4 \end{pmatrix}$$

$$\begin{pmatrix} 1 & 2 & 3 \\ 0 & -1 & -2 \\ 0 & 0 & 0 \end{pmatrix} \begin{pmatrix} x \\ y \\ z \end{pmatrix} = \begin{pmatrix} 4 \\ -6 \\ 2 \end{pmatrix}$$

$\Rightarrow$ third equation is $0 \times z = 2 \Rightarrow$ **no solution.**

The planes do not intersect: the planes are not parallel, but the line of intersection of any two is parallel to the third plane. The planes are said to form a prism.

(c)

$$\begin{pmatrix} 1 & 2 & 3 \\ 2 & 3 & 4 \\ 3 & 5 & 7 \end{pmatrix}\begin{pmatrix} x \\ y \\ z \end{pmatrix} = \begin{pmatrix} 4 \\ 2 \\ 6 \end{pmatrix} \qquad \begin{pmatrix} 1 & 2 & 3 \\ 0 & -1 & -2 \\ 0 & -1 & -2 \end{pmatrix}\begin{pmatrix} x \\ y \\ z \end{pmatrix} = \begin{pmatrix} 4 \\ -6 \\ -6 \end{pmatrix}$$

$$\begin{pmatrix} 1 & 2 & 3 \\ 0 & -1 & -2 \\ 0 & 0 & 0 \end{pmatrix}\begin{pmatrix} x \\ y \\ z \end{pmatrix} = \begin{pmatrix} 4 \\ -6 \\ 0 \end{pmatrix}$$

$\Rightarrow$ third equation is $0 \times z = 0 \Rightarrow z =$ any value.

Put $z = t \Rightarrow y = 6 - 2t$. From the second equation $y + 2z = 6$.
First equation is $x + 2y + 3z = 4 \Rightarrow x = 4 - 12 + 4t - 3t = t - 8$.
The solution is

$$\begin{pmatrix} x \\ y \\ z \end{pmatrix} = t\begin{pmatrix} 1 \\ -2 \\ 1 \end{pmatrix} + \begin{pmatrix} -8 \\ 6 \\ 0 \end{pmatrix}$$

which represents a straight line through the point $(-8,6,0)$ in the direction $\mathbf{i} - 2\mathbf{j} + \mathbf{k}$, i.e., the line joining the origin to $(1, -2.1)$. The three planes meet in a line, not a single point.

Example 29
Find the matrix X (in three dimensions) representing a half-turn about the x-axis and the matrix Y representing a half-turn about the y-axis. Find XY and YX and find the single transformation equivalent to X and Y performed successively.

For X $I(1,0,0)$ is invariant

$J(0,1,0) \rightarrow J'(0,-1,0)$ X is $\begin{pmatrix} 1 & 0 & 0 \\ 0 & -1 & 0 \\ 0 & 0 & -1 \end{pmatrix}$

$K(0,0,1) \rightarrow K'(0,0,-1)$

For Y $I(1,0,0) \rightarrow I'(-1,0,0)$

J is invariant Y is $\begin{pmatrix} -1 & 0 & 0 \\ 0 & 1 & 0 \\ 0 & 0 & -1 \end{pmatrix}$

$K(0,0,1) \rightarrow K'(0,0,-1)$

$$XY = \begin{pmatrix} 1 & 0 & 0 \\ 0 & -1 & 0 \\ 0 & 0 & -1 \end{pmatrix} \begin{pmatrix} -1 & 0 & 0 \\ 0 & 1 & 0 \\ 0 & 0 & -1 \end{pmatrix} = \begin{pmatrix} -1 & 0 & 0 \\ 0 & 1 & 0 \\ 0 & 0 & 1 \end{pmatrix}$$

$$Z$$

$$= \begin{pmatrix} -1 & 0 & 0 \\ 0 & -1 & 0 \\ 0 & 0 & -1 \end{pmatrix} \begin{pmatrix} 1 & 0 & 0 \\ 0 & 1 & 0 \\ 0 & 0 & -1 \end{pmatrix} = YX$$

The product matrix Z represents a half-turn about the z-axis. $XY = YX = Z$ which can be demonstrated in the following way. Hold a book in front of your face, front towards you, right way up. Imagine the origin at the centre of the book, the x-axis horizontal front to back, the y-axis side to side horizontal and the z-axis vertical top to bottom. (These axes form a right-handed set.) XY means Y first, X second, so you first turn the book away and end up, looking at the back upside down; then for X turn the book round (like the hand of a clock) to finish with the back the right way up; this is the result of doing Z.

YX means X first and Y second, so for X rotate (like a clock hand) to get front upside down, then do Y to get back right way up, i.e., Z. Students find three dimensional visualization difficult but it is well worthwhile trying to do this.

Try any combination of two of X, Y and Z. What is the result?

$ZYX = ZZ$ (since $YX = Z$) $= I$ since Z is its own inverse. Z is self-inverse $\Rightarrow Z^2 = I = X^2 = Y^2$.

Any combination of two of X, Y and Z produces the third.

Chapter 8

Series

This section of the syllabuses covers the various types of series and sequences and the methods for summing them. Most syllabuses mention the arithmetical and geometrical progressions, the binomial theorem and the expansions of trigonometrical, exponential, logarithmic and hyperbolic functions.

The use of Taylor's and Maclaurin's theorems, the method of induction and a limited knowledge of convergence (comparison and ratio tests) are required. The techniques required for selection, called permutations and combinations, are sometimes needed.

Most examination questions tend to test the principles by applying the methods to one or more of the series mentioned.

Multiple choice questions (Type A) Select the correct answer

Example 1

The sum of the first 10 terms of the series
$16+13+10+7+\ldots$ is

A 10   **B** 20   **C** 25   **D** 50   **E** 77

The series is an arithmetical progression since the difference of successive terms is constant. The common difference, d, equals -3.

The sum to n terms $S_n = \dfrac{n}{2}[2a+(n-1)d]$ where a is the first term, n is the number of terms and d is the common difference.

$$\therefore\ S_{10} = \tfrac{10}{2}[32+9(-3)] = 25 \qquad\qquad \text{ANSWER C}$$

Example 2

The sum to infinity of the series $0.03+0.003+0.0003+\ldots$ is

A $\tfrac{1}{3}$ **B** $\tfrac{1}{30}$ **C** $\tfrac{1}{33}$ **D** $\tfrac{10}{9}$ **E** infinity

---

This series is a geometrical progression with a common ratio of $r = \tfrac{1}{10}$ and a first term of $\tfrac{3}{100}$.

The sum to n terms $\quad S_n = \dfrac{a(1-r^n)}{1-r}$

However, since $-1 < r < 1$, $r^n \to 0$ as $n \to \infty$.

$$\therefore\ S_\infty = \frac{a}{1-r} = \tfrac{3}{100}\left(\frac{1}{1-\frac{1}{10}}\right) = \tfrac{3}{100} \times \tfrac{10}{9} = \tfrac{1}{30}$$

ANSWER B

Example 3

The range of validity of the expansion

$$\ln\left(\frac{1+x}{1-x}\right) = 2x + \frac{2x^3}{3} + \ldots + \frac{2x^{2n-1}}{(2n-1)} + \ldots \text{ is}$$

A $-1 < x \leqslant 1$ **B** $-1 \leqslant x < 1$ **C** $-1 \leqslant x \leqslant 1$
D $-1 < x < 1$ **E** all x

The given function is a combination of two logarithmic functions.

Thus, since $\ln\left(\dfrac{1+x}{1-x}\right) = \ln(1+x) - \ln(1-x)$ and $\ln(1+x)$ is valid

for $-1 < x \leqslant 1$ and $\ln(1-x)$ is valid for $-1 \leqslant x < 1$, the given expansion will only be valid when both of these conditions hold, i.e., when $-1 < x < 1$.   ANSWER D

Example 4

The first 3 terms of the expansion of $(9-2x)^{1/2}$ in ascending powers of x are

A $3 + \tfrac{1}{3}x + \tfrac{1}{54}x^2$ **B** $3 - \tfrac{1}{3}x + \tfrac{1}{54}x^2$ **C** $3 - \tfrac{1}{3}x - \tfrac{1}{54}x^2$
D $3 - \tfrac{1}{3}x + \tfrac{1}{12}x^2$ **E** $3 - \tfrac{1}{3}x - \tfrac{1}{12}x^2$

Now $\quad (9-2x)^{1/2} = 9^{1/2}(1 - \tfrac{2}{9}x)^{1/2} = 3(1 - \tfrac{2}{9}x)^{1/2}$

Using the binomial expansion for a rational index

$$3(1 - \tfrac{2}{9}x)^{1/2} = 3\left[1 - \tfrac{1}{2}(\tfrac{2}{9}x) + \frac{(\tfrac{1}{2})(-\tfrac{1}{2})}{2!}(-\tfrac{2}{9}x)^2 + \ldots\right]$$

$$= 3[1 - \tfrac{1}{9}x - \tfrac{1}{162}x^2 + \ldots] = 3 - \tfrac{1}{3}x - \tfrac{1}{54}x^2 + \ldots$$

ANSWER C

Example 5

The number of different arrangements that can be made by using all the letters of the word MINIMUM is

$$\textbf{A}\ 7! \quad \textbf{B}\ \frac{7!}{2} \quad \textbf{C}\ \frac{7!}{3} \quad \textbf{D}\ \frac{7!}{6} \quad \textbf{E}\ \frac{7!}{12}$$

If the letters were all different the number of arrangements would be 7!. However, in this case the letter M occurs three times and these can be arranged amongst themselves in 3! ways without altering the positions of the other letters.

Similarly, the letter I occurs twice and these can be arranged in 2! ways.

Hence the total number of distinct arrangements

$$= \frac{7!}{3!2!} = \frac{7!}{12}. \qquad\qquad \text{ANSWER E}$$

Example 6

The number of ways in which 3 books can be chosen from a shelf containing 8 different books is

$$\textbf{A}\ 8! \quad \textbf{B}\ 3! \quad \textbf{C}\ \frac{8!}{3!} \quad \textbf{D}\ \frac{8!}{5!} \quad \textbf{E}\ \frac{8!}{5!3!}$$

Since the order in which the 3 books are chosen is not important the question is asking for the number of different groups of 3 books that can be chosen, i.e., a combination. The number of ways 3 items can be chosen from 8 (where the order is important) is $\frac{8!}{5!}$. However, each group of 3 books can be rearranged amongst themselves in 3! ways.

Therefore the number of distinct groups of 3 books

$$= \frac{8!}{5!3!}. \qquad\qquad \text{ANSWER E}$$

Note that if you can remember that the number of ways of

choosing r items from n different items, where the order is not important, is ${}^nC_r = \dfrac{n!}{(n-r)!r!}$ then the solution can be found quickly.

Multiple choice questions (Type B) Answer according to the table

A	B	C	D	E
1, 2, 3	1, 3	2, 3	2	3
correct	only	only	only	only

Example 7
If $e^{x^2} = 1 + a_1 x + a_2 x^2 + \dots$

1 $a_r = 0$ for odd values of r **2** $a_2 = 1$

3 The general term is $\dfrac{x^{2r}}{r!}$

The expansion of $e^x = 1 + x + \dfrac{x^2}{2!} + \dfrac{x^3}{3!} + \dots$

Replacing x by x^2 we obtain

$$e^{x^2} = 1 + x^2 + \frac{x^4}{2!} + \frac{x^6}{3!} + \dots + \frac{x^{2r}}{r!} + \dots$$

By comparing with the given expansion it can be seen that $a_r = 0$ for all odd values of r, $a_2 = 1$ and the general term is as given. Statements 1, 2 and 3 are all true.   ANSWER A

Example 8
If $(1+x)^n = a_0 + a_1 x + a_2 x^2 + \dots + a_r x^r + \dots + a_n x^n$ where n is a positive integer.

1 $a_r = \dfrac{n(n-1)\dots(n-r-1)}{r!}$

2 $a_0 + a_1 + a_2 + \dots + a_n = 2^n$
3 Sum of the even coefficients = sum of the odd coefficients.

The binomial expansion for a positive integral index gives

$$(1+x)^n = 1 + {}^nC_1 x + {}^nC_2 x^2 + \dots + {}^nC_r x^r + \dots + x^n$$

$\therefore$ the coefficient of x^r is $a_r = {}^nC_r = \dfrac{n!}{(n-r)!\,r!}$

$$= \dfrac{n(n-1)(n-2)\ldots(n-r+1)}{r!}$$

Statement 1 is incorrect.

Substituting $x = 1$ in the expansion we obtain

$$a_0 + a_1 + a_2 + \ldots + a_n = 2^n$$

Statement 2 is correct.

Substituting $x = -1$

$$0 = a_0 - a_1 + a_2 - a_3 + \ldots + (-1)^n a_n$$

Hence, if n is even

$$a_1 + a_3 + a_5 + \ldots + a_{n-1} = a_0 + a_2 + a_4 + \ldots + a_n$$

If n is odd

$$a_1 + a_3 + a_5 + \ldots + a_n = a_0 + a_2 + a_4 + \ldots + a_{n-1}$$

In either case statement 3 is correct.
Statements 2 and 3 only are true. $\hfill$ ANSWER C

Multiple choice questions (Type C) For each question two statements are given. Answer

A if 1 implies 2 and 2 implies 1
B if 2 implies 1 but 1 does not imply 2
C if 1 implies 2 but 2 does not imply 1
D if 1 denies 2 and 2 denies 1
E if none of these relations holds

Example 9

1 $\displaystyle\sum_{r=1}^{n} a_r = 1 - \dfrac{1}{n+1}$ **2** $\displaystyle\sum_{r=1}^{\infty} a_r = 1$

From the result given in statement 1 it can be deduced that as

$n \to \infty$, the term $\dfrac{1}{n+1} \to 0$ and the sum to infinity $= 1$.

Thus statement 1 implies statement 2 but the reverse is not true. It is not possible to deduce the sum to n terms from the sum to infinity.

Statement 1 implies statement 2 but statement 2 does not imply statement 1. $\qquad$ ANSWER C

Example 10

1 The series Σu_n converges $\qquad$ **2** $\lim_{n \to \infty} u_n = 0$

Consider the series $S_n = \dfrac{1}{\sqrt{1}} + \dfrac{1}{\sqrt{2}} + \dfrac{1}{\sqrt{3}} + \ldots + \dfrac{1}{\sqrt{n}}$

Clearly the sum to n terms $S_n > n . \dfrac{1}{\sqrt{n}}$ since each term $> \dfrac{1}{\sqrt{n}}$.

$$\therefore S_n > \sqrt{n} \Rightarrow S_n \to \infty \text{ as } n \to \infty.$$

So although $u_n \to 0$ as $n \to \infty$ the series does not converge.

Conversely, if a series converges then $u_n \to 0$ as $n \to \infty$.
Since if $u_1 + u_2 + u_3 + \ldots$ is a convergent series then

$$u_n = S_n - S_{n-1} \Rightarrow \lim_{n \to \infty} u_n = \lim_{n \to \infty} S_n - \lim_{n \to \infty} S_{n-1} = 0$$

since S_n and S_{n-1} have the same limit as $n \to \infty$.

We say that the condition $\lim_{n \to \infty} u_n = 0$ is a necessary but not sufficient condition for convergence.

Statement 1 implies statement 2 but statement 2 does not imply statement 1. $\qquad$ ANSWER C

Remember the common example of this is the harmonic series

$1 + \tfrac{1}{2} + \tfrac{1}{3} + \tfrac{1}{4} + \ldots$ which diverges although $\lim_{n \to \infty} \dfrac{1}{n} = 0.$

Multiple choice questions (Type D) Each question consists of a problem followed by four pieces of information. Decide whether the problem can be solved with one of the four pieces of information omitted and answer

A if 1 could be omitted   **D** if 4 could be omitted
B if 2 could be omitted   **E** if none can be omitted
C if 3 could be omitted

Example 11
Find the first 6 terms in the expansion of the function $f(x)$ by using Maclaurin's theorem.

1 The curve passes through the origin
2 $f'(0) = 1$
3 $f''(0) + f(0) = -1$
4 $y_{n+2} + (2n+1)y_{n+1} + (n^2+1)y_n = 0$ where $y_n = f^n(0)$

The expansion of a function $f(x)$ by Maclaurin's theorem gives

$$f(x) = f(0) + xf'(0) + \frac{x^2}{2!}f''(0) + \frac{x^3}{3!}f'''(0) + \dots$$

Thus in order to determine the coefficient of each term it is necessary to evaluate successive differentials.

Statement 1 gives $f(0) = 0$ and statement 2 gives $f'(0) = 1$.
Statement 3 enables $f''(0)$ to be evaluated giving $f''(0) = -1$.
Statement 4 gives a recurrence relation between three consecutive coefficients and hence knowing the values of $f(0)$ and $f'(0)$ it is possible to evaluate $f''(0)$ and all the successive differentials.

$\therefore$ using statement 4 $f''(0) = -f'(0) - f(0) = -1$ $(n=0)$

Similarly when $n = 1$ $f'''(0) = -3f''(0) - 2f'(0) = 1$

and so on until all the required coefficients have been evaluated. Since this last statement enables all the coefficients to be calculated including $f''(0)$ the problem can be solved without statement 3.   ANSWER C

Note that, although the precise function has not been defined in this question, it is in fact given by $f(x) = \sin \ln (1+x)$.

Example 12
Find the sum to infinity of the series given by
$$u_0 + u_1 + u_2 + \dots + u_r + \dots$$

1 $u_r = ku_{r-1}$ where k is a constant   **2** $u_0 = 5$
3 $u_1 = 2$ **4** $-1 < k < 1$

Since, from statement 1, the ratio of two successive terms is a constant, k, the series is a geometrical progression which has k as its common ratio. Since $u_0 = 5$ and $u_1 = 2$, $k = \frac{2}{5}$. Thus $-1 < k < 1$ and the sum to infinity can be evaluated as

$$\frac{u_0}{1-k} = \frac{25}{3}$$

Statement 4 is not required. ANSWER D

Short questions

In this section the questions tend to require the expansions of one of the functions on the syllabus by the binomial theorem, Maclaurin's theorem or other standard method.

Example 13

If $(1 + \frac{3}{2}x)^{-1/2} = 1 - \frac{3}{4}x + ax^2 + bx^3$, as far as the term in x^3, find a and b.

Using the binomial theorem

$$(1 + \tfrac{3}{2}x)^{-1/2} = 1 + (-\tfrac{1}{2})(\tfrac{3}{2}x) + \frac{(-\tfrac{1}{2})(-\tfrac{3}{2})}{2!}(\tfrac{3}{2}x)^2$$

$$+ \frac{(-\tfrac{1}{2})(-\tfrac{3}{2})(-\tfrac{5}{2})}{3!}(\tfrac{3}{2}x)^3 + \dots$$

$$= 1 - \tfrac{3}{4}x + \tfrac{27}{32}x^2 - \tfrac{135}{128}x^3 + \dots$$

Hence $a = \frac{27}{32}$ and $b = -\frac{135}{128}$.

Example 14

Use the expansion of $(1 - x)^{1/2}$ to estimate the value of $\sqrt{11}$ to 4 decimal places.

Expanding by the binomial theorem

$$(1-x)^{1/2} = 1 - \tfrac{1}{2}x + \frac{(\tfrac{1}{2})(-\tfrac{1}{2})}{2!}(-x)^2 + \frac{(\tfrac{1}{2})(-\tfrac{1}{2})(-\tfrac{3}{2})}{3!}(-x)^3 + \dots$$

$$= 1 - \tfrac{1}{2}x - \tfrac{1}{8}x^2 - \tfrac{1}{16}x^3 - \dots$$

This is valid if $-1 < x < 1$. Let $x = 0.01$.

$$\therefore \ (1 - 0.01)^{1/2} = 1 - \tfrac{1}{2}(0.01) - \tfrac{1}{8}(0.01)^2 - \tfrac{1}{16}(0.01)^3 - \dots$$

$$\therefore \ \frac{\sqrt{99}}{10} \simeq 1 - 0.005 - 0.000\,012\,5 = 0.994\,987\,5$$

Since $\dfrac{\sqrt{99}}{10} = \dfrac{3\sqrt{11}}{10} \Rightarrow \sqrt{11} \simeq \frac{10}{3} \times 0.994\,987\,5 = 3.316\,625.$

$\therefore \sqrt{11} = 3.3166$ correct to 4 decimal places.

Example 15
Find the coefficient of x^n in the expansion of
$(a+bx)(1-2x)^{-1}$.

Expanding $(1-2x)^{-1}$ by the binomial theorem

$$(1-2x)^{-1} = 1 + 2x + \frac{(-1)(-2)}{2!}(-2x)^2$$

$$+ \frac{(-1)(-2)(-3)}{3!}(-2x)^3 + \dots$$

$\therefore \ (a+bx)(1-2x)^{-1} = (a+bx)[1 + 2x + (2x)^2 + \dots + (2x)^r + \dots]$

The term in x^n can be obtained by multiplying out the bracket choosing only the terms in x^{n-1} and x^n from the square bracket.

The term in $x^n = a(2x)^n + bx(2x)^{n-1} = 2^{n-1}(2a+b)x^n.$

The coefficient of the term in x^n is $(2a+b)2^{n-1}.$

Example 16
Expand $e^{\sin x}$ as far as the term in x^4 by using Maclaurin's theorem.

Maclaurin's theorem states

$$f(x) = f(0) + f'(0) + \frac{x^2}{2!}f''(0) + \frac{x^3}{3!}f'''(0) + \dots$$

If $f(x) = e^{\sin x} \Rightarrow f(0) = e^0 = 1$

$\therefore f'(x) = \cos x\, e^{\sin x} \Rightarrow f'(0) = 1$

$f''(x) = \cos^2 x\, e^{\sin x} - \sin x\, e^{\sin x} \Rightarrow f''(0) = 1$

$f'''(x) = \cos^3 x\, e^{\sin x} - 3\sin x \cos x\, e^{\sin x} - \cos x\, e^{\sin x}$

$\qquad = e^{\sin x}(\cos^3 x - \tfrac{3}{2}\sin 2x - \cos x) \Rightarrow f'''(0) = 0$

$f''''(x) = e^{\sin x}\{ -3\cos^2 x \sin x - 3\cos 2x - \cos x$

$\qquad\qquad + \cos^4 x - \tfrac{3}{2}\sin 2x \cos x - \cos^2 x\} \Rightarrow f''''(0) = -3$

Hence $e^{\sin x} = 1 + x + \dfrac{x^2}{2!} - \dfrac{3}{4!}x^4 + \dots$

$\qquad\qquad = 1 + x + \tfrac{1}{2}x^2 - \tfrac{1}{8}x^4$ as far as the term in x^4.

Alternatively, it would be possible to form a differential equation as follows. If $y = e^{\sin x}$ then $\dfrac{dy}{dx} = \cos x\, e^{\sin x} = y \cos x$ and $f'(0) = 1$.

Hence $\dfrac{d^2y}{dx^2} = -y \sin x + \cos x \dfrac{dy}{dx} \Rightarrow f''(0) = 1$

Also $\dfrac{d^3y}{dx^3} = -y \cos x - 2 \sin x \dfrac{dy}{dx} + \cos x \dfrac{d^2y}{dx^2} \Rightarrow f'''(0) = 0$

and so on until all the required coefficients have been found.

Example 17

Expand $\ln(3 + x)$ as a series in ascending powers of x giving the first 4 terms and the general term. State the range of values of x for which the expansion is valid.

$$\ln(3 + x) = \ln 3(1 + \tfrac{1}{3}x) = \ln 3 + \ln(1 + \tfrac{1}{3}x)$$

$$\therefore \ln(3 + x) = \ln 3 + \tfrac{1}{3}x - \tfrac{1}{2}(\tfrac{1}{3}x)^2 + \tfrac{1}{3}(\tfrac{1}{3}x)^3 - \dots + (-1)^{n-1}\frac{x^n}{n \cdot 3^n}$$

$$= \ln 3 + \tfrac{1}{3}x - \tfrac{1}{18}x^2 + \tfrac{1}{81}x^3 - \dots + (-1)^{n-1}\frac{x^n}{n \cdot 3^n}$$

The expansion is valid for $-1 < \tfrac{1}{3}x \leqslant 1$, i.e., $-3 < x \leqslant 3$.

Example 18

Expand $\sin^{-1} x$ as a series in ascending powers of x as far as the term in x^7.

Let $f(x) = \sin^{-1} x \Rightarrow f(0) = 0 \Rightarrow f'(x) = \dfrac{1}{\sqrt{1 - x^2}}$ $\qquad\qquad$ (1)

From (1) we have $(1 - x^2)^{1/2} f'(x) = 1$ and differentiating implicitly with respect to x

$$(1 - x^2)^{1/2} f''(x) - \frac{x}{(1 - x^2)^{1/2}} f'(x) = 0$$

or
$$(1-x^2)f''(x)-xf'(x)=0$$

By differentiating n times using Leibniz's theorem we obtain

$$(1-x^2)f^{n+2}(x)-2nxf^{n+1}(x)-n(n-1)f^n(x)-xf^{n+1}(x)+nf^n(x)=0$$

$$(1-x^2)f^{n+2}(x)-(2n+1)xf^{n+1}(x)-n^2f^n(x)=0$$

When $x=0$ this gives $f^{n+2}(0)=n^2f^n(0)$.

Now, since $f(0)=0$, $f^n(0)=0$ when n is even.

Also since $f'(0)=1 \Rightarrow f^3(0)=f'(0)=1^2$

$$\Rightarrow f^5(0)=3^2 \cdot f^3(0)=3^2 \cdot 1^2$$

$$\Rightarrow f^7(0)=5^2 \cdot f^5(0)=5^2 \cdot 3^2 \cdot 1^2$$

Hence using Maclaurin's theorem

$$\sin^{-1}x = x+\frac{1^2}{3!}x^3+\frac{3^2 \cdot 1^2}{5!}x^5+\frac{5^2 \cdot 3^2 \cdot 1^2}{7!}x^7+\dots$$

Notice that the general term of this series is given by

$$\frac{(2r-1)^2 \dots 5^2 \cdot 3^2 \cdot 1^2}{(2r+1)!}x^{2r+1} \text{ for } r \geq 1.$$

Long questions
Quite often full examination questions combine several shorter parts and may involve other topics such as partial fractions. One special technique is the difference method.

Example 19
Find the sum to n terms of the series

$$\frac{2}{1 \cdot 3 \cdot 5}+\frac{3}{3 \cdot 5 \cdot 7}+\frac{4}{5 \cdot 7 \cdot 9}+\dots$$

Deduce the sum to infinity.

The series is given by $S_n = \displaystyle\sum_{r=1}^{n} \frac{r+1}{(2r-1)(2r+1)(2r+3)}$

Consider

$$\frac{r+1}{(2r-1)(2r+1)(2r+3)} = \frac{A}{(2r-1)}+\frac{B}{(2r+1)}+\frac{C}{(2r+3)}$$

$$\equiv \frac{A(2r+1)(2r+3)+B(2r-1)(2r+3)+C(2r-1)(2r+1)}{(2r-1)(2r+1)(2r+3)}$$

Equating coefficients

$$r+1 \equiv A(2r+1)(2r+3) + B(2r-1)(2r+3) + C(2r-1)(2r+1)$$

Let $r = \tfrac{1}{2} \Rightarrow \tfrac{3}{2} = 8A \Rightarrow A = \tfrac{3}{16}$

$\quad r = -\tfrac{1}{2} \Rightarrow \tfrac{1}{2} = -4B \Rightarrow B = -\tfrac{1}{8}$

$\quad r = -\tfrac{3}{2} \Rightarrow -\tfrac{1}{2} = 8C \Rightarrow C = -\tfrac{1}{16}$

$$\therefore \sum_{r=1}^{n} \frac{r+1}{(2r-1)(2r+1)(2r+3)} = \tfrac{1}{16}\left(\frac{3}{2r-1} - \frac{2}{2r+1} - \frac{1}{2r+3}\right)$$

$$= \tfrac{1}{16}\Big(3 - \tfrac{2}{3} - \tfrac{1}{5}$$

$$\tfrac{3}{3} - \tfrac{2}{5} - \tfrac{1}{7}$$

$$\tfrac{3}{5} - \tfrac{2}{7} - \tfrac{1}{9}$$

$$\tfrac{3}{7} - \tfrac{2}{9} - \tfrac{1}{11}$$

$$\cdots\cdots\cdots\cdots$$

$$\cdots\cdots\cdots\cdots$$

$$\frac{3}{2n-5} - \frac{2}{2n-3} - \frac{1}{2n-1}$$

$$\frac{3}{2n-3} - \frac{2}{2n-1} - \frac{1}{2n+1}$$

$$\frac{3}{2n-1} - \frac{2}{2n+1} - \frac{1}{2n+3}\Big)$$

It can be seen that there are groups of three fractions which have a sum of zero as shown.

Hence $S_n = \tfrac{1}{16}\left(3 - \tfrac{2}{3} + \tfrac{3}{3} - \dfrac{1}{2n+1} - \dfrac{2}{2n+1} - \dfrac{1}{2n+3}\right)$

$$= \tfrac{1}{16}\left(\tfrac{10}{3} - \frac{3}{2n+1} - \frac{1}{2n+3}\right)$$

$$\therefore \sum_{r=1}^{n} \frac{r+1}{(2r-1)(2r+1)(2r+3)} = \frac{5}{24} - \frac{1}{16}\left(\frac{3}{2n+1} + \frac{1}{2n+3}\right)$$

Clearly as $n \to \infty$ the terms $\dfrac{3}{2n+1}$ and $\dfrac{1}{2n+3} \to 0$.

Hence $\displaystyle\sum_{r=1}^{\infty} \dfrac{r+1}{(2r-1)(2r+1)(2r+3)} = \dfrac{5}{24}$

Example 20

1. Find the number of ways a team of 4 can be chosen from 5 boys and 3 girls if
   (a) it must contain 2 boys and 2 girls,
   (b) it must contain at least 1 boy and 1 girl.

2. Find the sum to infinity of the series $\dfrac{2}{1!} + \dfrac{3}{2!} + \dfrac{4}{3!} + \ldots$

1(a) The boys can be chosen in $^5C_2$ ways $= \dfrac{5!}{3!2!}$ ways $= 10$ ways.

The girls can be chosen in $^3C_2$ ways $= \dfrac{3!}{2!1!}$ ways $= 3$ ways.

The total number of teams $= 10 \times 3 = 30$.

1(b) If the team must contain at least 1 boy and 1 girl it can be formed in the following ways.

1 boy and 3 girls $= {}^5C_1 \times {}^3C_3 = 5 \times 1 = 5$ ways

2 boys and 2 girls $= 30$ ways   (from part (a))

3 boys and 1 girl $= {}^5C_3 \times {}^3C_1 = 10 \times 3 = 30$ ways
The total number of teams $= 5 + 30 + 30 = 65$.

2. The series $\dfrac{2}{1!} + \dfrac{3}{2!} + \dfrac{4}{3!} + \ldots$ can be written $\displaystyle\sum_{r=1}^{\infty} \dfrac{r+1}{r!}$

But $\displaystyle\sum_{r=1}^{\infty} \dfrac{r+1}{r!} = \sum_{r=1}^{\infty} \left(\dfrac{1}{(r-1)!} + \dfrac{1}{r!} \right)$

Consider $\quad e^x = 1 + x + \dfrac{x^2}{2!} + \dfrac{x^3}{3!} + \ldots + \dfrac{x^r}{r!} + \ldots$

when $x = 1 \quad e = 1 + 1 + \dfrac{1}{2!} + \dfrac{1}{3!} + \ldots + \dfrac{1}{r!} + \ldots$

i.e. $\displaystyle\sum_{r=1}^{\infty} \dfrac{1}{r!} = e - 1$ and $\displaystyle\sum_{r=1}^{\infty} \dfrac{1}{(r-1)!} = e$

$\therefore \displaystyle\sum_{r=1}^{\infty} \dfrac{r+1}{r!} = e + (e-1) = 2e - 1$

Chapter 9

Complex Numbers

The invention of a complex number was due to Gauss and it enabled mathematicians to proceed further in finding roots of an equation of any degree. Knowledge of the algebraic form, $x+iy$, the polar form, $r(\cos\theta+i\sin\theta)$ and the exponential form are expected by most of the syllabuses.

Representation of complex numbers on an Argand diagram and use of the diagram are also required.

One important result is de Moivre's theorem although the extent to which it is included depends upon the syllabus. Its application to roots of unity, trigonometrical identities, integration and differentiation may be necessary for those studying further. Loci and transformations in the complex plane may also be required.

Multiple choice questions (Type A) Select the correct answer

Example 1

The roots of the equation $x^2+2x+4=0$ are

A $\frac{1}{2}(-2\pm3\sqrt{2}i)$ **B** $\frac{1}{2}(2\pm3\sqrt{2}i)$ **C** $-1\pm\sqrt{3}i$

D $1\pm\sqrt{3}i$ **E** none

Using the formula solution for the equation $ax^2+bx+c=0$

$$x=\frac{-b\pm\sqrt{b^2-4ac}}{2a} \qquad \Rightarrow x=\frac{-2\pm\sqrt{(-2)^2-4(1)(4)}}{2}$$

$$\therefore x=\frac{-2\pm\sqrt{-12}}{2}=\frac{-2\pm2\sqrt{3}i}{2}=-1\pm\sqrt{3}i \qquad \text{ANSWER C}$$

Note that answer E is not correct for, although there are no real roots, complex roots do exist.

Example 2

$$\frac{3+2i}{2-i} =$$

A $\dfrac{4+7i}{3}$ **B** $\dfrac{4+7i}{5}$ **C** $\dfrac{8+i}{3}$ **D** $\dfrac{8+i}{5}$ **E** $\dfrac{8-i}{5}$.

Multiply numerator and denominator by the complex conjugate of the denominator, i.e., by $2+i$:

$$\frac{(3+2i)(2+i)}{(2-i)(2+i)} = \frac{6+3i+4i+2i^2}{4-i^2} = \frac{4+7i}{5} \qquad \text{ANSWER B}$$

Example 3

$$(3-2i)^2 =$$

A $5-6i$ **B** $13-12i$ **C** $13-6i$ **D** $5-12i$ **E** 5

$$(3-2i)^2 = (3-2i)(3-2i) = 9-6i-6i+4i^2 = 5-12i$$

$$\text{ANSWER D}$$

Example 4

The modulus-argument (polar) form of $-2+2\sqrt{3}i$ is

A $4(\cos\frac{2\pi}{3}+i\sin\frac{2\pi}{3})$ **B** $4(\cos\frac{5\pi}{6}+i\sin\frac{5\pi}{6})$
C $-4(\cos\frac{\pi}{3}-i\sin\frac{\pi}{3})$ **D** $\frac{1}{4}(\cos\frac{2\pi}{3}+i\sin\frac{2\pi}{3})$
E $\frac{1}{4}(\cos\frac{5\pi}{6}+i\sin\frac{5\pi}{6})$

Now $-2+2\sqrt{3}i = 4(-\frac{1}{2}+\frac{1}{2}\sqrt{3}i) = 4(\cos\frac{2\pi}{3}+i\sin\frac{2\pi}{3})$

$$\text{ANSWER A}$$

Remember that the form must be $r(\cos\theta+i\sin\theta)$. Therefore $1-\sqrt{3}i = 2(\frac{1}{2}-\frac{1}{2}\sqrt{3}i) = 2[\cos(-\frac{\pi}{3})+i\sin(-\frac{\pi}{3})]$ and not $2[\cos\frac{\pi}{3}-i\sin\frac{\pi}{3}]$. The argument can easily be checked on an Argand diagram.

Example 5

In Figure 22 the points P_1 and P_2 represent the complex numbers z_1 and z_2. The point P_3 represents the complex number

A z_1-z_2 **B** z_1+z_2 **C** z_2-z_1 **D** z_1z_2 **E** z_1/z_2

The sum of two complex numbers z_1 and z_2, represented by the points P_1 and P_2, is found geometrically as the fourth vertex, R, of a parallelogram OP_1P_2R.

Since

$$z_1 - z_2 = z_1 + (-z_2)$$

it follows that, since P_4 represents $(-z_2)$, P_3 represents $z_1 - z_2$ as it is the fourth vertex of the completed parallelogram $OP_1P_4P_3$.

ANSWER A

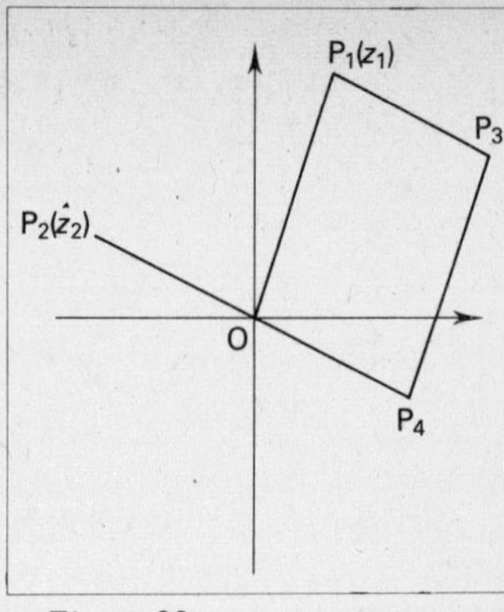

Figure 22

Example 6
The circle shown in Figure 23 is given by the equation

A $|z-2-3i| = 2$ **B** $|z+2-3i| = 4$ **C** $|z-2+3i| = 2$
D $|z-2+3i| = 4$ **E** $|z+2-3i| = 2$

In general, the distance of a variable point (representing z) from a fixed point (representing a) in the Argand plane is $|z-a|$. If this distance is constant the locus is a circle.

In this case

$$|z-(2-3i)| = 2$$

or $|z-2+3i| = 2$.

ANSWER C

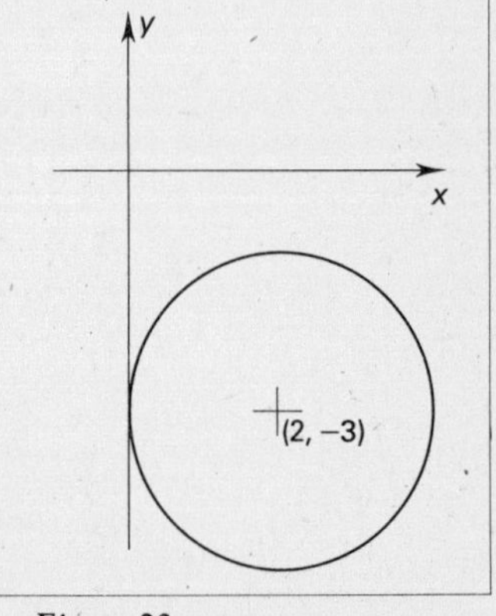

Figure 23

Alternatively, the cartesian equation can be formed by letting $z = x+iy$. For answer C this would produce

$$|x+iy-2+3i| = 2 \Rightarrow |x-2+i(y+3)| = 2.$$

By the definition of modulus, $(x-2)^2 + (y+3)^2 = 4$ which is a circle, centre $(2,-3)$ and radius 2.

Multiple choice questions (Type B) Answer according to the table

A	B	C	D	E
1, 2, 3	1, 3	2, 3	2	3
correct	only	only	only	only

Example 7

If $z_1 = 1+i$ and $z_2 = 1-\sqrt{3}i$ are two complex numbers

1 $|z_1 z_2| = \dfrac{1}{\sqrt{2}}$ **2** $\arg(z_1 z_2) = -\tfrac{\pi}{12}$

3 $|z_1 - z_2| = 1+\sqrt{3}$

Changing z_1 and z_2 into polar form

$$z_1 = 1+i = \sqrt{2}\left(\frac{1}{\sqrt{2}}+\frac{1}{\sqrt{2}}i\right) = \sqrt{2}(\cos\tfrac{\pi}{4}+i\sin\tfrac{\pi}{4})$$

$$z_2 = 1-\sqrt{3}i = 2\left(\frac{1}{2}-\frac{\sqrt{3}}{2}i\right) = 2[\cos(-\tfrac{\pi}{3})+i\sin(-\tfrac{\pi}{3})]$$

Hence $z_1 z_2 = 2\sqrt{2}\,[\cos(\tfrac{\pi}{4}-\tfrac{\pi}{3})+i\sin(\tfrac{\pi}{4}-\tfrac{\pi}{3})]$ since the product of z_1 and z_2 is obtained by taking the product of the moduli and the sum of the arguments.

$$\therefore\ z_1 z_2 = 2\sqrt{2}\,[\cos(-\tfrac{\pi}{12})+i\sin(-\tfrac{\pi}{12})]$$

Statement 1 is incorrect since the modulus is $2\sqrt{2}$.
Statement 2 is correct as the argument is $-\tfrac{\pi}{12}$.
Now $z_1 - z_2 = (1+\sqrt{3})i \Rightarrow |z_1 - z_2| = 1+\sqrt{3}$ and statement 3 is correct.

Statements 2 and 3 only are true.   ANSWER C

Example 8

If $z = x+iy$ is a variable complex number
1 $z\bar{z} = x^2 + y^2$
2 The locus defined by $z\bar{z} = k$ where k is a constant is a circle
3 The maximum value of $|z|$ such that $|z-(1+i)| \leqslant 1$ is $1+\sqrt{2}$

If $z = x+iy$ then its complex conjugate $\bar{z} = x-iy$.
Hence $z\bar{z} = (x+iy)(x-iy) = x^2+y^2$

Statement 1 is correct.

The locus defined by $z\bar{z} = k$ is equivalent to $x^2+y^2 = k$ which represents a circle, centre $(0,0)$ and radius $\sqrt{k}$.

Statement 2 is correct.

If $|z-(1+i)| \leqslant 1$, then the distance between z and the point representing $1+i$ is $\leqslant 1$, i.e., the point representing z lies on or within the circle, centre $(1,1)$ and radius 1. The maximum value of $|z|$ is obtained when this point is furthest from the origin.

Therefore the maximum value of z is $1+\sqrt{2}$, since the centre of the circle $(1,1)$ is a distance $\sqrt{2}$ from the origin. The total distance is therefore $\sqrt{2}+$ (the radius of the circle). Notice that the minimum value of z will be $\sqrt{2}-1$.

Statement 3 is correct.

Statements 1, 2 and 3 are all true. ANSWER A

Multiple choice questions (Type C) For each question two statements are given. Answer

A if 1 implies 2 and 2 implies 1
B if 2 implies 1 but 1 does not imply 2
C if 1 implies 2 but 2 does not imply 1
D if 1 denies 2 and 2 denies 1
E if none of these relations holds

Example 9

1 $z = -1+i$ 2 $z = \sqrt{2}\left(\cos\frac{3\pi}{4}+i\sin\frac{3\pi}{4}\right)$

$z = -1+i$ can be written as $\sqrt{2}\left(-\dfrac{1}{\sqrt{2}}+\dfrac{1}{\sqrt{2}}i\right)$

$\therefore z = -1+i = \sqrt{2}\left[\cos\frac{3\pi}{4}+i\sin\frac{3\pi}{4}\right]$

Statement 1 implies statement 2.

Conversely, $\sqrt{2}(\cos\frac{3\pi}{4}+i\sin\frac{3\pi}{4})=\sqrt{2}(-\cos\frac{\pi}{4}+i\sin\frac{\pi}{4})$

$$=\sqrt{2}\left[-\frac{1}{\sqrt{2}}+\frac{1}{\sqrt{2}}i\right]=-1+i$$

Statement 2 implies statement 1.
Statement 1 implies statement 2 and statement 2 implies statement 1. ANSWER A

Example 10

z_1 and z_2 are two complex numbers.

1 $\arg z_1+\arg z_2=0$ **2** z_1 and z_2 are complex conjugates

---

If $\arg z_1=\alpha$ then $\arg z_2=-\alpha$ since $\arg z_1+\arg z_2=0$.

However, two complex numbers with arguments of α and $-\alpha$ are not necessarily complex conjugates as $|z_1|$ may not equal $|z_2|$.
Statement 1 does not imply statement 2.
If $z_1=x+iy$ and $z_2=x-iy$ (i.e., complex conjugates) then

$\arg z_1=\tan^{-1}\left(\dfrac{y}{x}\right)=\alpha$ and $\arg z_2=\tan^{-1}\left(-\dfrac{y}{x}\right)=-\alpha$ (taking the

principal values).

Hence $\qquad\qquad \arg z_1+\arg z_2=\alpha-\alpha=0$
Statement 2 implies statement 1.
Statement 2 implies statement 1 but statement 1 does not imply statement 2. ANSWER B

Short questions
On many examination papers the questions on complex numbers consist of several parts testing different aspects of the topic. The following examples illustrate these different types.

Example 11

If $z_1=2+3i$ and $z_2=2-i$ express

(a) z_1-2z_2 (b) $z_1^2+z_2^2$ (c) z_1/z_2 in the form $a+bi$

(a) $z_1 - 2z_2 = 2 + 3i - 2(2 - i) = -2 + 5i$

(b) $z_1^2 + z_2^2 = (2 + 3i)^2 + (2 - i)^2$
$$= 4 + 12i + 3i^2 + 4 - 4i + i^2 = 4 + 8i$$

(c) $\dfrac{z_1}{z_2} = \dfrac{2 + 3i}{2 - i} = \dfrac{(2 + 3i)(2 + i)}{(2 - i)(2 + i)} = \dfrac{4 + 8i + 3i^2}{5} = \dfrac{1 + 8i}{5}$

Example 12
Find the square roots of $1 + \sqrt{3}i$.

Now $1 + \sqrt{3}i = 2\left(\dfrac{1}{2} + \dfrac{\sqrt{3}}{2}i\right) = 2[\cos(\tfrac{\pi}{3} + 2k\pi) + i\sin(\tfrac{\pi}{3} + 2k\pi)]$

Using De Moivre's theorem

$(1 + \sqrt{3}i)^{1/2} = 2^{1/2}[\cos(\tfrac{\pi}{3} + 2k\pi) + i\sin(\tfrac{\pi}{3} + 2k\pi)]^{1/2}$
$$= 2^{1/2}[\cos(\tfrac{\pi}{6} + k\pi) + i\sin(\tfrac{\pi}{6} + k\pi)]$$

If $k = 0$, the root is $\sqrt{2}(\cos\tfrac{\pi}{6} + i\sin\tfrac{\pi}{6}) = \dfrac{1}{\sqrt{2}}(\sqrt{3} + i)$.

If $k = 1$ the root is $\sqrt{2}(\cos\tfrac{7\pi}{6} + i\sin\tfrac{7\pi}{6}) = -\dfrac{1}{\sqrt{2}}(\sqrt{3} + i)$.

The square roots of $1 + \sqrt{3}i$ are $\pm\dfrac{1}{\sqrt{2}}(\sqrt{3} + i)$.

Alternatively, this can be solved by assuming that the root is of the form $a + bi$.

$\therefore\ (a + bi)^2 = 1 + \sqrt{3}i \Rightarrow a^2 - b^2 = 1$ and $2ab = \sqrt{3}$.

Solving $\qquad a^2 - \dfrac{3}{4a^2} = 1 \Rightarrow 4a^4 - 4a^2 - 3 = 0$

$\qquad\qquad (2a^2 + 1)(2a^2 - 3) = 0 \Rightarrow a^2 = \tfrac{3}{2}$ or $-\tfrac{1}{2}$

Since a is real $\quad a = \pm\sqrt{\tfrac{3}{2}} \quad b = \pm\sqrt{\tfrac{1}{2}}$

The square roots are $\pm\dfrac{1}{\sqrt{2}}(\sqrt{3} + i)$.

Example 13

If P is the point representing the complex number z on the Argand diagram and $|z| = 2|z+3i|$ find the equation of the locus of P. Sketch the locus of P.

Let $z = x+iy$ and since $|z| = 2|z+3i| \Rightarrow |z|^2 = 4|z+3i|^2$

we have
$$x^2 + y^2 = 4[x^2 + (y+3)^2]$$

$$x^2 + y^2 = 4x^2 + 4y^2 + 24y + 36$$

$$3x^2 + 3y^2 + 24y + 36 = 0 \Rightarrow x^2 + y^2 + 8y + 12 = 0$$

This can be written as $x^2 + (y+4)^2 = 4$ which represents a circle, centre $(0, -4)$ and radius 2. The sketch is shown in Figure 24(a). This is a particular example of the general result which states that if A and B are two fixed points, the locus of a point P, which moves such that $PA:PB$ is constant, is a circle.

The circle is called the circle of Apollonius.

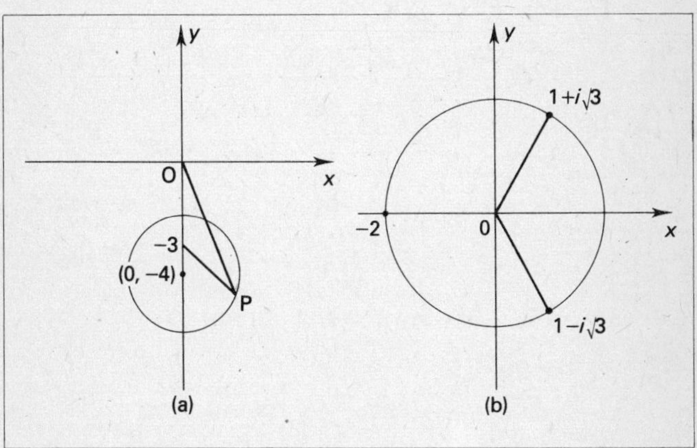

Figure 24

Example 14

Solve the equation $z^3 + 8 = 0$ and show the roots on an Argand diagram.

If $f(z) = z^3 + 8$ then $f(-2) = 0$ and hence by the factor theorem $z + 2$ is a factor.

Now $z^3 + 8 = (z+2)(z^2 - 2z + 4) = 0$

$\therefore z = -2$ or $z^2 - 2z + 4 = 0$

If $z^2 - 2z + 4 = 0$ then $z = \dfrac{2 \pm \sqrt{4 - 16}}{2} = 1 \pm i\sqrt{3}$

Hence the roots are $-2,\ 1 \pm i\sqrt{3}$ which are shown in Figure 24(b).

Notice that the roots lie on a circle, centre (0,0) and radius 2 and are equally spaced around the circle. In general the use of De Moivre's theorem provides an alternative method which may be necessary for equations of higher degree.

Example 15

Use De Moivre's theorem to express $\sin^5\theta$ in terms of the sum of the cosines of multiples of θ.

Let $z = \cos\theta + i\sin\theta \Rightarrow \dfrac{1}{z} = z^{-1} = \cos\theta - i\sin\theta$

So $z^n = (\cos\theta + i\sin\theta)^n = \cos n\theta + i\sin n\theta$ and

$$\frac{1}{z^n} = (\cos\theta - i\sin\theta)^n = \cos n\theta - i\sin n\theta$$

Hence $\quad (2i\sin\theta)^5 = \left(z - \dfrac{1}{z}\right)^5$

$$32i\sin^5\theta = \left(z^5 - \frac{1}{z^5}\right) - 5\left(z^3 - \frac{1}{z^3}\right) + 10\left(z - \frac{1}{z}\right)$$

$$= 2i(\sin 5\theta - 5\sin 3\theta + 10\sin\theta)$$

$$\therefore \quad \sin^5\theta = \tfrac{1}{16}(\sin 5\theta - 5\sin 3\theta + 10\sin\theta)$$

Example 16

Use De Moivre's theorem to express $\sin 5\theta / \sin\theta$ as the sum of powers of $\cos\theta$.

By De Moivre's theorem $\cos 5\theta + i\sin 5\theta = (\cos\theta + i\sin\theta)^5$

Hence $\cos 5\theta + i\sin 5\theta = \cos^5\theta + 5i\cos^4\theta\sin\theta$

$$+ 10i^2 \cos^3\theta \sin^2\theta + 10i^3 \cos^2\theta \sin^3\theta + 5i^4 \cos\theta \sin^4\theta + i^5 \sin^5\theta$$

Equating imaginary parts

$$\sin 5\theta = 5\cos^4\theta \sin\theta - 10\cos^2\theta \sin^3\theta + \sin^5\theta$$

$$\therefore \quad \frac{\sin 5\theta}{\sin\theta} = 5\cos^4\theta - 10\cos^2\theta \sin^2\theta + \sin^4\theta$$

$$= 5\cos^4\theta - 10\cos^2\theta(1-\cos^2\theta) + (1-\cos^2\theta)^2$$

$$= 16\cos^4\theta - 12\cos^2\theta + 1$$

Example 17

Express $e^{i\pi/4}$ in the form $a+bi$.

Since $e^{i\theta} = \cos\theta + i\sin\theta$ $\qquad e^{\frac{i\pi}{4}} = \cos\frac{\pi}{4} + i\sin\frac{\pi}{4}$

Thus $e^{\frac{i\pi}{4}} = \dfrac{1}{\sqrt{2}} + \dfrac{1}{\sqrt{2}}i = \dfrac{1}{\sqrt{2}}(1+i)$

Long questions

The following examples of full questions illustrate how the different aspects of the work on complex numbers can be woven together to develop an idea or used as separate and unrelated parts.

Example 18

The complex numbers $z_1 = \dfrac{\sqrt{2}(1-3i)}{2-i}$ and $z_2 = (1+i)^2$ are represented by the points P_1 and P_2 respectively on the Argand diagram.

(i) Express z_1 and z_2 in the form $a+bi$.

(ii) Show that OP_1P_2 is an isosceles triangle if O is the origin.

(iii) Find the locus of the point P representing the complex number z if $|z-z_1| = |z-z_2|$.

(iv) Describe the locus of the point Q which represents the complex number z if $\arg\left(\dfrac{z-z_1}{z-z_2}\right) = \pm\dfrac{\pi}{2}$.

(v) Sketch the two loci on the same Argand plane.

(i) $z_1 = \dfrac{\sqrt{2}(1-3i)}{2-i} = \dfrac{\sqrt{2}(1-3i)(2+i)}{(2-i)(2+i)} = \dfrac{\sqrt{2}(2-5i-3i^2)}{5}$

$$= \frac{\sqrt{2}(5-5i)}{5} = \sqrt{2}(1-i)$$

$$z_2 = (1+i)^2 = 1+2i+i^2 = 2i$$

(ii) The points P_1 and P_2 are shown in Figure 25(a).

Since $|z_1| = |\sqrt{2}-\sqrt{2}i| = \sqrt{(\sqrt{2})^2+(\sqrt{2})^2} = \sqrt{4} = 2$

and $|z_2| = 2$

it follows that P_1 and P_2 are equidistant from the origin and hence OP_1P_2 is an isosceles triangle.

(iii) If $z = x+iy$ then the locus $|z-z_1| = |z-z_2|$ gives

$$|(x-\sqrt{2})+i(y+\sqrt{2})| = |x+i(y-2)|$$

$$(x-\sqrt{2})^2+(y+\sqrt{2})^2 = x^2+(y-2)^2$$

$$(2\sqrt{2}+4)y = 2\sqrt{2}x \Rightarrow y = \frac{\sqrt{2}x}{2+\sqrt{2}}$$

Rationalizing the denominator

$$y = \frac{\sqrt{2}x(2-\sqrt{2})}{(2+\sqrt{2})(2-\sqrt{2})} = (\sqrt{2}-1)x$$

The equation of the locus of P is $y = (\sqrt{2}-1)x$.

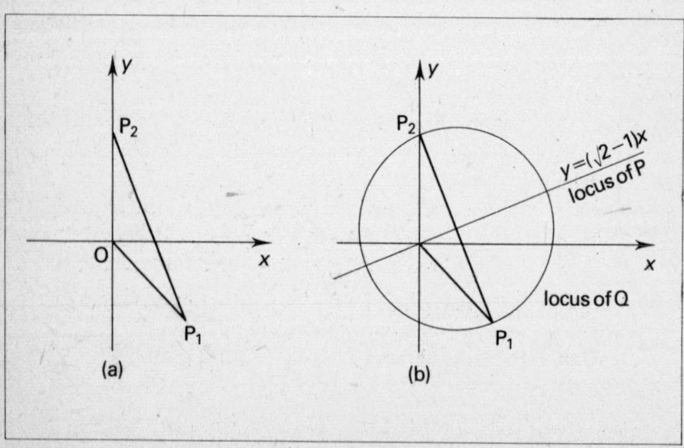

Figure 25

(iv) Since $\arg\left(\dfrac{z-z_1}{z-z_2}\right) = \arg(z-z_1) - \arg(z-z_2)$

$$\arg(z-z_1) - \arg(z-z_2) = \pm\tfrac{\pi}{2}$$

This indicates that the angle between the lines joining Q to P_1 and P_2 is $90°$. Hence Q moves on a circle with P_1P_2 as diameter.

(v) The sketch is shown in Figure 25(b).

Example 19

   (i) Given that one root of the equation $z^3 + az^2 + bz - 15 = 0$ is $2+i$, find the values of a and b and solve the equation if a and b are real.

   (ii) Express $e^{x(5+3i)}$ in the form $e^{5x}(\cos ax + i\sin ax)$. Hence find $\int e^{5x}\sin 3x\, dx$.

   (iii) Express $\cos ix$ and $\sin ix$ in terms of $\cosh x$ and $\sinh x$ respectively. If $u + iv = \tan^{-1}(x+iy)$ show that

$$\tanh 2v = \frac{2y}{1+x^2+y^2}.$$

Since the coefficients of the equation are real any complex roots must occur in complex conjugate pairs. Thus if $2+i$ is a root so is $2-i$.

Hence $[z-(2+i)]$ and $[z-(2-i)]$ are factors.

Since $[z-(2+i)][z-(2-i)] = z^2 - 4z + 5$ we have
$$z^3 + az^2 + bz - 15 = (z^2 - 4z + 5)(z-p) = 0$$
where the real root is $z = p$.

Equating the coefficients we obtain

constant term	$-5p = -15 \Rightarrow p =$	3
coefficient of z^2	$-4-p = a \quad \Rightarrow a = -7$	
coefficient of z	$4p+5 = b \quad \Rightarrow b =$	17

The values required are $a = -7$, $b = 17$ and the roots of the equation $z^3 - 7z^2 + 17z - 15 = 0$ are $z = 3$, $z = 2\pm i$.

(ii) Now $e^{x(5+3i)} = e^{5x}\cdot e^{3ix}$

Since $e^{in\theta} = \cos n\theta + i\sin n\theta$

$$e^{x(5+3i)} = e^{5x}(\cos 3x + i\sin 3x)$$

$$\therefore \int e^{5x}(\cos 3x + i \sin 3x)\,dx = \int e^{(5+3i)x}\,dx$$

$$= \frac{1}{5+3i}e^{(5+3i)x} = \frac{(5-3i)}{34}e^{5x}(\cos 3x + i \sin 3x)$$

Equating the imaginary parts

$$\int e^{5x}\sin 3x\,dx = \frac{1}{34}e^{5x}[5\sin 3x - 3\cos 3x] + c$$

(iii) Since $\qquad\qquad e^{i\theta} = \cos\theta + i\sin\theta$

and $\qquad\qquad e^{-i\theta} = \cos\theta - i\sin\theta$

it follows that $\cos\theta = \tfrac{1}{2}(e^{i\theta} + e^{-i\theta})$ and $\sin\theta = \dfrac{1}{2i}(e^{i\theta} - e^{-i\theta})$

$$\therefore\ \cos ix = \tfrac{1}{2}(e^{i(ix)} + e^{-i(ix)}) = \tfrac{1}{2}(e^{x} + e^{-x}) = \cosh x \qquad\qquad (1)$$

$$\sin ix = \frac{1}{2i}(e^{i(ix)} - e^{-i(ix)}) = -\frac{1}{2i}(e^{x} - e^{-x}) = i\sinh x \qquad\qquad (2)$$

Now if $u + iv = \tan^{-1}(x+iy)$ then $u - iv = \tan^{-1}(x-iy)$

Subtracting, $2iv = \tan^{-1}(x+iy) - \tan^{-1}(x-iy)$

$$\therefore\ \tan 2iv = \tan[\tan^{-1}(x+iy) - \tan^{-1}(x-iy)]$$

$$= \frac{(x+iy) - (x-iy)}{1 + (x+iy)(x-iy)}$$

$$= \frac{2iy}{1 + x^2 + y^2}$$

But dividing (2) by (1) we have $\tan ix = i\tanh x$

$$\therefore \qquad\qquad \tanh 2v = \frac{2y}{1 + x^2 + y^2}.$$

Chapter 10
Differential Equations

Differential equations contain at least one differential coefficient. The study of such equations and their solutions is extensive but syllabuses restrict the range to first and second order equations. The techniques for solving include separation of the variable, linear equations requiring an integrating factor, homogeneous equations and the methods needed for linear second order equations with constant coefficients. Numerical solutions by the step-by-step process and the use of Taylor's theorem may also be required.

Multiple choice questions (Type A) Select the correct answer

Example 1

Given that $y = 1$ when $x = 1$ the solution of the equation

$$x^2 \frac{dy}{dx} = y \text{ is}$$

A $\ln y = \dfrac{1}{x} - 1$ **B** $\ln y = -\dfrac{1}{x}$ **C** $\ln y = 1 - \dfrac{1}{x}$

D $\ln y = -\dfrac{2}{x^3}$ **E** $\ln y = 2 - \dfrac{2}{x}$

Separating the variable $\qquad x^2 \dfrac{dy}{dx} = y \Rightarrow \dfrac{1}{y} \dfrac{dy}{dx} = \dfrac{1}{x^2}$

Integrating with respect to x

$$\int \frac{1}{y} \frac{dy}{dx} \, dx = \int \frac{1}{x^2} \, dx \Rightarrow \int \frac{1}{y} \, dy = \int \frac{1}{x^2} \, dx$$

$$\therefore \ \ln y = -\frac{1}{x} + c$$

$y = 1$ when $x = 1 \Rightarrow c = 1.$

The particular solution is $\quad \ln y = 1 - \dfrac{1}{x}.$ $\qquad$ ANSWER C

Example 2

The general solution of the equation $y \dfrac{dy}{dx} + x = 1$ defines a family of

A circles, centre (0,0) $\qquad$ **B** circles, centre (1,0)
C parabolas, vertex (1,0) $\qquad$ **D** parabolas, vertex (0,0)
E ellipses, centre (0,0)

In this equation the variables can be separated to give

$$y\frac{dy}{dx} = 1 - x$$

Integrating with respect to x

$$\int y\frac{dy}{dx}\,dx = \int 1 - x\,dx \Rightarrow \int y\,dy = \int 1 - x\,dx$$

$$\therefore \qquad \tfrac{1}{2}y^2 = x - \tfrac{1}{2}x^2 + B \Rightarrow y^2 + x^2 - 2x = 2B$$

or $(x-1)^2 + y^2 = C$ where $C = 2B + 1$.

This general solution represents a family of circles with centres at (1,0). ANSWER B

Example 3
A suitable integrating factor for converting the equation
$\dfrac{dy}{dx} - \dfrac{2y}{x} = x^2$ into an exact equation is

   A e^{-x^2} **B** e^{-x} **C** $-x^2$ **D** $\dfrac{2}{x^2}$ **E** $\dfrac{1}{x^2}$

The first order linear equation $\dfrac{dy}{dx} + Py = Q$, where P and Q are functions of x only, can be rearranged by multiplying by the integrating factor defined by $e^{\int P\,dx}$.

In this case the integrating factor is $e^{\int -\frac{2}{x}dx}$

$$= e^{-2\ln x} = e^{\ln x^{-2}} = x^{-2} = \frac{1}{x^2} \qquad\qquad \text{ANSWER E}$$

Note that if we multiply by this factor we obtain

$$\frac{1}{x^2}\frac{dy}{dx} - \frac{2}{x^3}y = 1 \Rightarrow \frac{d}{dx}\left(\frac{y}{x^3}\right) = 1$$

and the solution is given by $\dfrac{y}{x^3} = x + c$, i.e., $y = x^4 + cx^3$

Candidates may be asked to formulate a differential equation given readings from a given situation. Examples taken from other subjects will not assume a knowledge of that subject.

136

Example 4

A liquid is being heated in such a way that its temperature rise is proportional to the inverse of the time for which the heat is applied. At the same time it loses heat at a rate proportional to the amount by which its temperature exceeds room temperature. If θ is its temperature at time t and room temperature is taken as 20°C then the differential equation describing this model is

A $\dfrac{d\theta}{dt} - a\theta = -20a + \dfrac{b}{t}$ **B** $\dfrac{d\theta}{dt} + a\theta = 20a + \dfrac{b}{t}$

C $\dfrac{d\theta}{dt} + a\theta = 20a + bt$ **D** $\dfrac{d\theta}{dt} - a\theta = -20a + bt$

E $\dfrac{d\theta}{dt} + a\theta = -20a + \dfrac{b}{t}$

If θ is the temperature of the liquid at time t, the rate of change of temperature is $\dfrac{d\theta}{dt}$. The gain is proportional to the inverse of time

$\left(\text{i.e. } \dfrac{b}{t}\right)$ although it also loses heat at a rate proportional to its temperature above 20°C (i.e. $-a(\theta - 20)$).

$\therefore \dfrac{d\theta}{dt} = \dfrac{b}{t} - a(\theta - 20)$ or $\dfrac{d\theta}{dt} + a\theta = 20a + \dfrac{b}{t}$. **ANSWER B**

Example 5

If $x = 2$ and $\dfrac{dx}{dt} = 18$ when $t = 0$ the particular solution of

the equation $\dfrac{d^2x}{dt^2} + 3\dfrac{dx}{dt} - 10x = 0$ is given by

$x = Ae^{2t} + Be^{-5t}$ where

A $A = 0, B = 2$ **B** $A = 4, B = -2$ **C** $A = -\frac{8}{3}, B = \frac{14}{3}$
D $A = \frac{20}{3}, B = -\frac{14}{3}$ **E** $A = \frac{14}{3}, B = -\frac{8}{3}$

The general solution is given as $x = Ae^{2t} + Be^{-5t}$ (1)

This gives $\dfrac{dx}{dt} = 2Ae^{2t} - 5Be^{-5t}$ (2)

It is possible to find the constants A and B by substituting the boundary conditions into equations (1) and (2).

$x = 2$ when $t = 0 \Rightarrow A + B = 2$

$\dfrac{dx}{dt} = 18$ when $t = 0 \Rightarrow 2A - 5B = 18$

Solving gives $A = 4$ and $B = -2$. $\hspace{2cm}$ **ANSWER B**

Remember that a second order equation has two arbitrary constants in its general solution and therefore requires two boundary conditions if these are to be evaluated.

Multiple choice questions (Type B) Answer according to the table

A	B	C	D	E
1, 2, 3	1, 3	2, 3	2	3
correct	only	only	only	only

Example 6

The motion of a particle is given by the equation $\dfrac{d^2x}{dt^2} = -\omega^2 x$ where x is the displacement of the particle at time t and ω is a constant.

1 The velocity, v, is given by $v^2 = \omega^2(a^2 - x^2)$ where $x = a$ when $v = 0$.
2 The solution is $x = a \cos(\omega t + \alpha)$ where α is a constant.
3 The motion is periodic.

Let $v = \dfrac{dx}{dt} \Rightarrow \dfrac{d^2x}{dt^2} = \dfrac{dv}{dt} = \dfrac{dv}{dx} \cdot \dfrac{dx}{dt} = v \dfrac{dv}{dx}$

Hence the given equation becomes $v \dfrac{dv}{dx} = -\omega^2 x$.

Integrating with respect to x

$$\int v \, dv = -\int \omega^2 x \, dx \Rightarrow \tfrac{1}{2}v^2 = -\tfrac{1}{2}\omega^2 x^2 + c$$

If $x = a$ when $v = 0$ we have $c = \tfrac{1}{2}\omega^2 a^2$

$$\therefore \qquad v^2 = \omega^2(a^2 - x^2) \qquad\qquad (1)$$

Statement 1 is correct.

Now from (1) $\qquad -\dfrac{dx}{dt} = \pm\omega(a^2 - x^2)^{1/2} \qquad\qquad (2)$

Separating the variable and integrating with respect to t

$$-\int \frac{1}{\sqrt{a^2 - x^2}} \frac{dx}{dt} \, dt = \pm \int \omega \, dt \Rightarrow -\int \frac{1}{\sqrt{a^2 - x^2}} \, dx = \pm \int \omega \, dt$$

$\therefore \cos^{-1}\left(\dfrac{x}{a}\right) = \pm(\omega t + \alpha)$ where α is a constant.

$\therefore x = a\cos[\pm(\omega t + \alpha)] \Rightarrow x = a\cos(\omega t + \alpha)$.

Statement 2 is correct.

Since the displacement is given by $x = a\cos(\omega t + \alpha)$, if we let $x = a$ when $t = 0$ then $\cos\alpha = 1 \Rightarrow \alpha = 0$.

Hence a particular solution is $x = a\cos\omega t$ and the particle will pass through a point of displacement x_1 $(-a \leqslant x_1 \leqslant a)$ for values of t differing by 2π.
Thus the motion is periodic and statement 3 is correct.

Statements 1, 2 and 3 are all true. **ANSWER A**

Note that the choice of the minus sign in Equation (2) avoids an ambiguity of sign in the later working.

This equation defines the motion of a particle performing simple harmonic oscillations.

Example 7

The motion of a particle is defined by the equation $\dfrac{d^2x}{dt^2} + 2\dfrac{dx}{dt} + 5x = 0$ where $x = 5$ and $\dfrac{dx}{dt} = -5$ when $t = 0$.

1 $x = 5e^{-t}\cos 2t$ is a particular solution.
2 The motion is oscillatory.
3 Successive amplitudes form a geometrical progression.

As $x = 5e^{-t}\cos 2t$, we have, differentiating with respect to t

$$\frac{dx}{dt} = 5e^{-t}(-2\sin 2t) - 5e^{-t}\cos 2t = 5e^{-t}(-2\sin 2t - \cos 2t)$$

$$\frac{d^2x}{dt^2} = 5e^{-t}(-4\cos 2t + 2\sin 2t) - 5e^{-t}(-2\sin 2t - \cos 2t)$$

Substituting in the equation we obtain

$$\frac{d^2x}{dt^2} + 2\frac{dx}{dt} + 5x = 5e^{-t}(-4\cos 2t + 2\sin 2t + 2\sin 2t + \cos 2t$$
$$-4\sin 2t - 2\cos 2t + 5\cos 2t)$$

Hence $\dfrac{d^2x}{dt^2} + 2\dfrac{dx}{dt} + 5x = 0$ and thus $x = 5e^{-t}\cos 2t$ is a particular solution.

Statement 1 is correct.

The motion is oscillatory since when $x = 0$, say, $\cos 2t = 0$ giving $t = (2n-1)\frac{\pi}{4}$ for $n = 1, 2, 3, \ldots$.

Clearly the particle passes through the point $x = 0$ at successive times separated by an interval of $\dfrac{\pi}{2}$ seconds.

Statement 2 is correct.

The maximum amplitudes occur when $\cos 2t = \pm 1$

i.e., when $2t = 0, \pi, 2\pi, \ldots$ $t = 0, \dfrac{\pi}{2}, \pi, \dfrac{3\pi}{2}, \ldots$.

The maximum amplitudes are $x_1 = 5$, $x_2 = -5e^{-\frac{\pi}{2}}$, $x_3 = 5e^{-\pi} \ldots$.

These form a geometrical progression with common ratio $-e^{-\frac{\pi}{2}}$ and thus statement 3 is correct.

Statements 1, 2 and 3 are all true. **ANSWER A**

Multiple choice questions (Type C) For each question two statements are given. Answer

A if 1 implies 2 and 2 implies 1
B if 2 implies 1 but 1 does not imply 2
C if 1 implies 2 but 2 does not imply 1
D if 1 denies 2 and 2 denies 1
E if none of these relations holds

Example 8

1 $y = e^{-x} + e^{-3x} + 2 \sin x - \tfrac{3}{2} \cos x$

2 $\dfrac{d^2y}{dx^2} + 4\dfrac{dy}{dx} + 3y = 5 \cos x + 10 \sin x$

If $y = e^{-x} + e^{-3x} + 2 \sin x - \tfrac{3}{2} \cos x$

then $\dfrac{dy}{dx} = -e^{-x} - 3e^{-3x} + 2 \cos x + \tfrac{3}{2} \sin x$

and $\dfrac{d^2y}{dx^2} = e^{-x} + 9e^{-3x} - 2 \sin x + \tfrac{3}{2} \cos x$.

$\therefore \dfrac{d^2y}{dx^2} + 4\dfrac{dy}{dx} + 3y = (e^{-x} + 9e^{-3x} - 2 \sin x + \tfrac{3}{2} \cos x)$

$$+ 4(-e^{-x} - 3e^{-3x} + 2 \cos x + \tfrac{3}{2} \sin x)$$

$$+ 3(e^{-x} + e^{-3x} + 2 \sin x - \tfrac{3}{2} \cos x)$$

$$\therefore \quad \frac{d^2y}{dx^2} + 4\frac{dy}{dx} + 3y = 5\cos x + 10\sin x.$$

Hence y is a particular solution of the equation and statement 1 implies statement 2.

The general solution of a second order differential equation requires two arbitrary constants. No boundary conditions are given and statement 2 does not imply statement 1.   **ANSWER C**

Example 9

The motion of a particle of unit mass moving in a straight line Ox under the action of a force n^2x directed towards O and a resisting force $2k\dfrac{dx}{dt}$ is given by the equation

$\dfrac{d^2x}{dt^2} + 2k\dfrac{dx}{dt} + n^2x = 0$ where x is the distance from O, and n and k are constants.

1 $x = Ce^{-kt}\cos(\omega t + \varepsilon)$ where $\omega^2 = n^2 - k^2$ and ε is a constant.

2 $k^2 < n^2$

The given differential equation is linear with constant coefficients. The auxiliary equation will be

$$p^2 + 2kp + n^2 = 0 \Rightarrow p = \frac{-2k \pm \sqrt{4k^2 - 4n^2}}{2} = -k \pm \sqrt{k^2 - n^2}$$

Hence if $\omega^2 = n^2 - k^2$ the general solution is

$$x = e^{-kt}(Ae^{i\omega t} + Be^{-i\omega t})$$

$$\therefore \quad x = e^{-kt}[A(\cos\omega t + i\sin\omega t) + B(\cos\omega t - i\sin\omega t)]$$

$$= e^{-kt}[(A+B)\cos\omega t + i(A-B)\sin\omega t]$$

$$= e^{-kt}(D\cos\omega t + E\sin\omega t)$$

$$= Ce^{-kt}\cos(\omega t + \varepsilon) \text{ with } C^2 = D^2 + E^2 \text{ and } \tan\varepsilon = -\frac{E}{D}.$$

Hence statement 2 implies statement 1. Also statement 1 implies statement 2 since if the solution is $x = Ce^{-kt}\cos(\omega t + \varepsilon)$ the condition $k^2 < n^2$ must apply.   **ANSWER A**

Note that the solution of $\dfrac{d^2x}{dt^2} + 2k\dfrac{dx}{dt} + n^2x = 0$ will depend upon the nature of the roots of the auxiliary equation

$$p^2 + 2kp + n^2 = 0$$

(a) If $k^2 > n^2$, real distinct roots are obtained and the solution is $x = e^{-kt}[Ae^{\sqrt{(k^2-n^2)}t} + Be^{-\sqrt{(k^2-n^2)}t}]$. Since $x \neq 0$ for finite t the motion is not oscillatory although $x \to 0$ as $t \to \infty$.

(b) If $k^2 = n^2$, the roots are repeated and the solution is written as $x = e^{-kt}(A + Bt)$. Again this is not oscillatory since $x = 0$ for only one value of t.

(c) If $k^2 < n^2$ the roots are complex and the solution is formed as above. This gives damped harmonic motion.

Short questions

The examples in this section illustrate the various methods required for solving differential equations. Many examination questions test these techniques.

Example 10

Solve the differential equation $\dfrac{dy}{dx} + \sec^2 y = 0$.

Rearranging and integrating with respect to x this becomes

$$\int \cos^2 y \, \frac{dy}{dx} \, dx = -\int dx \Rightarrow \int \cos^2 y \, dy = -\int dx$$

Hence
$$\tfrac{1}{2}\int (1 + \cos 2y) \, dy = -x + c$$

$$\therefore \quad \frac{y}{2} + \frac{\sin 2y}{4} = -x + c$$

The general solution is $x = \tfrac{1}{4}(4c - 2y - \sin 2y)$.

Example 11

Solve the differential equation $\dfrac{1}{x}\dfrac{dy}{dx} - y^2 = 1$.

This is an equation where the variables can be separated.

Hence
$$\frac{1}{x}\frac{dy}{dx} = y^2 + 1 \Rightarrow \frac{1}{y^2 + 1}\frac{dy}{dx} = x$$

Integrating with respect to x

$$\int \frac{1}{y^2 + 1} \, dy = \int x \, dx \Rightarrow \tan^{-1} y = \tfrac{1}{2}x^2 + c$$

> **Example 12**
>
> Solve the equation $\dfrac{dy}{dx} + 3y = e^{-3x} \cos x$.

This equation is linear and of the form $\dfrac{dy}{dx} + Py = Q$ where P and Q are functions of x only. It is solved by using the integrating factor $e^{\int P\,dx}$. In this case the integrating factor is $e^{\int 3\,dx} = e^{3x}$.

Multiplying by this factor the equation becomes

$$e^{3x}\frac{dy}{dx} + 3e^{3x}y = e^{3x}\cdot e^{-3x}\cos x \Rightarrow \frac{d}{dx}(e^{3x}y) = \cos x$$

Integrating with respect to x

$$e^{3x}y = \sin x + C \quad \text{or} \quad y = e^{-3x}\sin x + Ce^{-3x}$$

> **Example 13**
>
> Solve the equation $\dfrac{dy}{dx} + y\tan x = \sec x$, given that $y = \dfrac{1}{\sqrt{2}}$ when $x = \frac{\pi}{4}$.

Again this is a linear equation and we use the integrating factor $e^{\int P\,dx}$.

In this case the factor is $e^{\int \tan x\,dx} = e^{-\ln \cos x} = \sec x$. Multiplying by this factor the equation becomes

$$\sec x\,\frac{dy}{dx} + y\sec x\tan x = \sec^2 x \Rightarrow \frac{d}{dx}(y\sec x) = \sec^2 x$$

Integrating with respect to x we obtain the general solution

$$y\sec x = \tan x + C$$

If $y = \dfrac{1}{\sqrt{2}}$ when $x = \frac{\pi}{4}$ then $\dfrac{1}{\sqrt{2}}\sec\frac{\pi}{4} = \tan\frac{\pi}{4} + C$.

This gives $C = 0$ and the particular solution is

$$y\sec x = \tan x \quad \text{or} \quad y = \sin x$$

Note that the integrating factor is always of the form $e^{\int P\,dx}$ but this may simplify to give a much easier expression as in Example 13. Remember that $e^{\ln x} = x$.

Example 14

Solve the equation $x^2 \dfrac{dy}{dx} = 4x^2 + 5xy + y^2$.

Since the equation is of the form $P\dfrac{dy}{dx} = Q$ where P and Q are homogeneous functions of x and y of the same degree we divide by x^2 to give

$$\frac{dy}{dx} = 4 + \frac{5y}{x} + \left(\frac{y}{x}\right)^2$$

Let $\dfrac{y}{x} = v \Rightarrow y = vx \Rightarrow \dfrac{dy}{dx} = v + x\dfrac{dv}{dx}$

Substituting, the equation becomes

$$v + x\frac{dv}{dx} = 4 + 5v + v^2 \quad \text{or} \quad x\frac{dv}{dx} = 4 + 4v + v^2$$

Hence
$$\frac{1}{(2+v)^2}\frac{dv}{dx} = \frac{1}{x}$$

Integrating with respect to x

$$\int (2+v)^{-2}\, dv = \int \frac{1}{x}\, dx \Rightarrow -(2+v)^{-1} = \ln kx$$

But $v = \dfrac{y}{x}$ and thus

$$-\left(2 + \frac{y}{x}\right)^{-1} = \ln kx \quad \text{or} \quad -\frac{x}{(2x+y)} = \ln kx$$

Second order differential equations include equations of the form $\dfrac{d^2y}{dx^2} = f(y)$, equations containing $\dfrac{d^2y}{dx^2}$ and $\dfrac{dy}{dx}$ but no term in x or y, and the most general linear equation with constant coefficients $a\dfrac{d^2y}{dx^2} + b\dfrac{dy}{dx} + cy = f(x)$. The particular integrals needed in the solution of this last equation will be limited to a few specific cases which can be found by trial.

Example 15

Solve the equation $(1 - x^2)\dfrac{d^2y}{dx^2} + 2x\dfrac{dy}{dx} = 0$.

Let $p = \dfrac{dy}{dx} \Rightarrow \dfrac{d^2y}{dx^2} = \dfrac{dp}{dx}$

The equation becomes $(1-x^2)\dfrac{dp}{dx} = -2xp$

Separating the variables and integrating with respect to x

$$\int \frac{1}{p}\,dp = \int \frac{-2x}{(1-x^2)}\,dx$$

$\therefore\ \ln p = \ln k(1-x^2) \Rightarrow p = \dfrac{dy}{dx} = k(1-x^2)$

Integrating this result with respect to x we obtain the general solution

$$y = kx - \tfrac{1}{3}kx^3 + C$$

Note that if the equation contains the first and second derivatives but no term in x then the same method is used except that $\dfrac{d^2y}{dx^2}$ is written as $p\dfrac{dp}{dy}$.

The general second order linear equation with constant coefficients

$$a\frac{d^2y}{dx^2} + b\frac{dy}{dx} + cy = f(x)$$

can be solved easily when $f(x) = 0$. The format of the solution depends upon the nature of the roots of the auxiliary equation.

<hr>

Example 16

Solve the equation $\dfrac{d^2y}{dx^2} - 5\dfrac{dy}{dx} + 4y = 0.$

<hr>

Assuming a solution of the form $y = Ae^{kx}$ the auxiliary equation will be $k^2 - 5k + 4 = 0.$

Thus $(k-4)(k-1) = 0 \Rightarrow k = 1$ or $k = 4.$

Hence $y = Ae^x$ and $y = Be^{4x}$ are particular solutions and the complete general solution is

$$y = Ae^x + Be^{4x}$$

<hr>

Example 17

Solve the equation $\dfrac{d^2y}{dx^2} - 10\dfrac{dy}{dx} + 25y = 0.$

<hr>

The auxiliary equation is $k^2 - 10k + 25 = 0$ giving
$(k-5)^2 = 0 \Rightarrow k = 5$ twice.

This only produces one particular solution ($y = Ce^{5x}$). Since the general solution needs two arbitrary constants we write the solution in the form

$$y = (A + Bx)e^{5x}$$

Example 18

Solve the equation $\dfrac{d^2y}{dx^2} - 4\dfrac{dy}{dx} + 8y = 0$.

The auxiliary equation is $k^2 - 4k + 8 = 0$ giving (by applying the formula)

$$k = \frac{4 \pm \sqrt{16 - 32}}{2} = 2 \pm 2i \quad \text{where} \quad i^2 = -1.$$

The general solution is $y = Ae^{(2+2i)x} + Be^{(2-2i)x}$

$$= e^{2x}(Ae^{2ix} + Be^{-2ix})$$

Using $e^{2ix} = \cos 2x + i \sin 2x$ and $e^{-2ix} = \cos 2x - i \sin 2x$ we have, after writing $C = A + B$ and $D = i(A - B)$,

$$y = e^{2x}(C \cos 2x + D \sin 2x)$$

The full linear equation will be satisfied by $y = y_1 + y_2$ where y_1 is the solution of the reduced equation obtained by putting $f(x) = 0$ and y_2 is a particular solution found by trial.

Example 19

Solve $\dfrac{d^2y}{dx^2} - 5\dfrac{dy}{dx} + 4y = 4x^2 - 2x - 3$.

The reduced equation $\dfrac{d^2y}{dx^2} - 5\dfrac{dy}{dx} + 4y = 0$ has a general solution $y_1 = Ae^x + Be^{4x}$ (see Example 16).

To find a particular integral when $f(x)$ is a polynomial try a polynomial of the same degree.

Let $y_2 = a + bx + cx^2$. Substituting in the given equation we obtain

$$2c - 5(b + 2cx) + 4(a + bx + cx^2) = 4x^2 - 2x - 3$$

Equating coefficients

x^2 term $\qquad\qquad\qquad 4c = 4 \Rightarrow c = 1$

x term $\qquad\qquad -10c + 4b = -2 \Rightarrow 4b = 8 \Rightarrow b = 2$

constant term $\qquad\qquad 2c - 5b + 4a = -3$

and by rearranging we obtain $4a = 5b - 2c - 3$

$\therefore \ 4a = 5(2) - 2(1) - 3 \Rightarrow 4a = 5 \Rightarrow a = \frac{5}{4}.$

A particular solution is $y_2 = x^2 + 2x + \frac{5}{4}.$

The complete general solution is

$$y = Ae^x + Be^{4x} + x^2 + 2x + \tfrac{5}{4}$$

Remember that the form of the trial solution depends upon the nature of $f(x)$. If $f(x)$ is of the form Ae^{ax} then try $y_2 = Ce^{ax}$ but if e^{ax} is also a solution of the reduced equation then try $y_2 = Cxe^{ax}$.

If $f(x)$ is of the form $A\cos ax + B\sin ax$ then try a particular solution $y_2 = C\cos ax + D\sin ax$.

There are many differential equations for which exact solutions cannot be found. In such cases a numerical approximation process can be used which produces values that nearly lie on the solution curve.

A step by step method takes a starting point on the curve and evaluates a neighbouring point by assuming that the curve approximates to the tangent to the curve using the linear approximation $\delta y \simeq f'(x)\delta x$.

A better approximation can be achieved by using a quadratic approximation, of the form

$$\delta y \simeq f'(x)\delta x + \tfrac{1}{2}f''(x)(\delta x)^2$$

Chapter 11

Vectors

Questions involving vectors are many and varied, ranging from short questions on the algebra of vectors to geometrical applications and the uses of vector methods in applied mathematics. Different examination boards use different notations and it is wise to be thoroughly conversant with the notation used in each examination paper.

Multiple choice questions (Type A) Select the correct answer

Example 1

P divides AB in the ratio $1:2$. Referred to the origin O, $OP =$

A $\mathbf{OA}+2\mathbf{OB}$ B $2\mathbf{OA}+\mathbf{OB}$ C $\frac{2}{3}\mathbf{OA}+\frac{1}{3}\mathbf{OB}$

D $\frac{1}{3}\mathbf{OA}+\frac{2}{3}\mathbf{OB}$ E $2\mathbf{OA}-\mathbf{OB}$

$$\mathbf{OP} = \mathbf{OA}+\mathbf{AP} = \mathbf{OA}+\tfrac{1}{3}\mathbf{AB}$$

$$= \mathbf{OA}+\tfrac{1}{3}(\mathbf{AO}+\mathbf{OB})$$

$$= \mathbf{OA}-\tfrac{1}{3}\mathbf{OA}+\tfrac{1}{3}\mathbf{OB}$$

$$= \tfrac{2}{3}\mathbf{OA}+\tfrac{1}{3}\mathbf{OB}$$

ANSWER C

This is the ratio theorem which can be quoted. If the ratio is $1:2$ the fractions are $\frac{1}{3}$ and $\frac{2}{3}$. To make sure these are the right way round ($\frac{2}{3}$ with $\mathbf{OA}$, $\frac{1}{3}$ with $\mathbf{OB}$) remember that P is nearer A and will contain more $\mathbf{OA}$ than $\mathbf{OB}$.

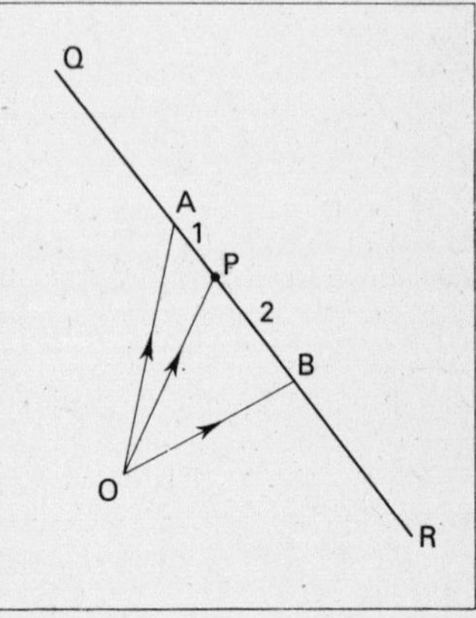

Figure 26

Example 2

Q divides AB in the ratio $-1:2$. $\mathbf{OQ} =$

A $-\mathbf{OA}+2\mathbf{OB}$ B $\mathbf{OA}-2\mathbf{OB}$ C $\frac{1}{3}\mathbf{OA}-\frac{2}{3}\mathbf{OB}$

D $\frac{2}{3}\mathbf{OA}-\frac{1}{3}\mathbf{OB}$ E $2\mathbf{OA}-\mathbf{OB}$

With a negative ratio Q is either on AB produced or BA produced. $AQ:QB = -1:2 \Rightarrow$ the distance AQ compared with QB is 1:2 in magnitude and opposite in direction. This can happen only when Q is outside A (on BA produced) as shown in Figure 26. It is easier to regard A as the mid-point of Q and B and apply the ratio theorem to get

$$\mathbf{OA} = \tfrac{1}{2}\mathbf{OQ} + \tfrac{1}{2}\mathbf{OB} \Rightarrow 2\mathbf{OA} = \mathbf{OQ} + \mathbf{OB} \Rightarrow \mathbf{OQ} = 2\mathbf{OA} - \mathbf{OB}$$

ANSWER E

If R divides AB in the ratio $-2:1$ (or $2:-1$), $AR:RB = -2:1$ and R is outside B. In this case $\mathbf{OR} = 2\mathbf{OB} - \mathbf{OA}$ and as R is nearer to B than A there is more of $\mathbf{OB}$ than $\mathbf{OA}$ in the composition of $\mathbf{OR}$.

If P (or Q) is anywhere on AB, the ratios add up to 1.

Example 3

The direction cosines of the vector $3\mathbf{i} - 4\mathbf{j} + 12\mathbf{k}$ are

A $3, -4, 12$ **B** $3, 4, 12$ **C** $\frac{3}{19}, -\frac{4}{19}, \frac{12}{19}$ **D** $\frac{3}{11}, -\frac{4}{11}, \frac{12}{11}$
E $\frac{3}{13}, -\frac{4}{13}, \frac{12}{13}$

The direction cosines of a vector are the cosines of the 3 angles that the vector makes with the positive directions of the three co-ordinate axes Ox, Oy and Oz.

If $\mathbf{OP} = 3\mathbf{i} - 4\mathbf{j} + 12\mathbf{k}$ and a line is drawn from P to meet Ox at right angles at X, then the angle POx is given by $\cos POx = \frac{3}{13}$ since the length OP is $\sqrt{3^2 + 4^2 + 12^2} = 13$.

For the angles OP makes with Oy and Oz a similar result holds to give $\cos POy = -\frac{4}{13}$ and $\cos POz = \frac{12}{13}$.   ANSWER E

The direction of a vector can be specified by the angles which it makes with the three coordinate axes. This method is the easiest way to find these three angles and since it involves finding the cosines of the angles, the three ratios are called the direction cosines.

Example 4

The plane $x - 2y + 3z = 6$ meets the coordinate axes Ox, Oy and Oz

A at the origin   **B** at (1,0,0)   **C** at (3,0,0)   **D** at (6,0,0)
        (0,−2,0)   (0,2,0)   (0,3,0)
E at (6,0,0)   (0,0,3)   (0,0,1)   (0,0,2)
      (0,−3,0)
      (0,0,2)

If the plane meets Ox at X then X is given by $y = 0$ and $z = 0 \Rightarrow x = 6$. Similarly $-2y = 6 \Rightarrow y = -3$ at $Y(0, -3, 0)$ and Z is $(0,0,2)$. E is the correct answer. **ANSWER E**

Example 5
The scalar product of the vectors $\mathbf{i} + 2\mathbf{j} + 3\mathbf{k}$ and $3\mathbf{i} - 2\mathbf{j} + \mathbf{k}$ is

A $3\mathbf{i} + 4\mathbf{j} + 3\mathbf{k}$ **B** $3\mathbf{i} - 4\mathbf{j} + 3\mathbf{k}$ **C** 10 **D** 2 **E** 36

The scalar product of two vectors is a number (a scalar is a number). The respective components are multiplied together and then added.

$$\mathbf{a} \cdot \mathbf{b} = (\mathbf{i} + 2\mathbf{j} + 3\mathbf{k}) \cdot (3\mathbf{i} - 2\mathbf{j} + \mathbf{k}) = 3 - 4 + 3 = 2 \qquad \text{ANSWER D}$$

Example 6
The vector product of the vectors $\mathbf{a} = \mathbf{i} + 2\mathbf{j} + 3\mathbf{k}$ and $\mathbf{b} = 3\mathbf{i} - 2\mathbf{j} + \mathbf{k}$ is

A $3\mathbf{i} + 4\mathbf{j} + 3\mathbf{k}$ **B** $3\mathbf{i} - 4\mathbf{j} + 3\mathbf{k}$ **C** $8\mathbf{i} - 8\mathbf{j} - 8\mathbf{k}$
D $8\mathbf{i} + 8\mathbf{j} - 8\mathbf{k}$ **E** $\mathbf{i} + \mathbf{j} - \mathbf{k}$

$$\mathbf{a} \times \mathbf{b} = \begin{vmatrix} \mathbf{i} & \mathbf{j} & \mathbf{k} \\ 1 & 2 & 3 \\ 3 & -2 & 1 \end{vmatrix} = \mathbf{i}(+2+6) - \mathbf{j}(1-9) + \mathbf{k}(-2-6)$$
$$= 8\mathbf{i} + 8\mathbf{j} - 8\mathbf{k} \qquad \text{ANSWER D}$$

The usual mistake is to get the sign of the middle component $(\mathbf{j})$ wrong. As a check $\mathbf{a} \times \mathbf{b}$ is perpendicular to both $\mathbf{a}$ and $\mathbf{b}$.

Multiple choice questions (Type B) Answer according to the table

A	B	C	D	E
1, 2, 3	1, 3	2, 3	2	3
correct	only	only	only	only

Example 7
The vector equation of AB is

1 $\mathbf{r} = \mathbf{OA} + t\mathbf{AB}$ **2** $\mathbf{r} = \mathbf{OA} + s\mathbf{BA}$ **3** $\mathbf{r} = \mathbf{OB} + p\mathbf{BA}$

The vector equation of a straight line is not unique: there are several possible forms. Three are given in this question. If the equation is $\mathbf{r} = \mathbf{OC} + t\mathbf{EF}$, C must be a point on the line and the vector $\mathbf{EF}$ must represent the direction of the line. All three answers are correct. ANSWER A

Example 8
In Figure 27, $\mathbf{OD} =$

1 $\frac{1}{3}\mathbf{OA} + \frac{1}{3}\mathbf{OB} + \frac{1}{3}\mathbf{OC}$ **2** $\frac{5}{9}\mathbf{OA} + \frac{4}{9}\mathbf{OC} + \frac{1}{3}\mathbf{OB}$
3 $\frac{2}{3}(\frac{1}{2}\mathbf{OA} + \frac{1}{2}\mathbf{OB}) + \frac{1}{3}\mathbf{OC}$

By coordinates

$$\mathbf{OA} + \mathbf{OB} + \mathbf{OC} = 6\mathbf{j} + 6\mathbf{i} + 9\mathbf{i} + 6\mathbf{j}$$
$$= 15\mathbf{i} + 12\mathbf{j}$$
$$\tfrac{1}{3}(\mathbf{OA} + \mathbf{OB} + \mathbf{OC}) = 5\mathbf{i} + 4\mathbf{j}$$

Statement 1 is correct.

$$\tfrac{5}{9}\mathbf{OA} + \tfrac{4}{9}\mathbf{OC} + \tfrac{1}{3}\mathbf{OB} = 3\tfrac{1}{3}\mathbf{j} + 4\mathbf{i} + 2\tfrac{2}{3}\mathbf{j} + 2\mathbf{i}$$
$$= 6\mathbf{i} + 6\mathbf{j}$$

Statement 2 is not correct.
Statement 3 is the same as statement 1, but suggests another method.
$\mathbf{OM} = \frac{1}{2}\mathbf{OA} + \frac{1}{2}\mathbf{OB}$ and D divides CM in the ratio $2:1$ so

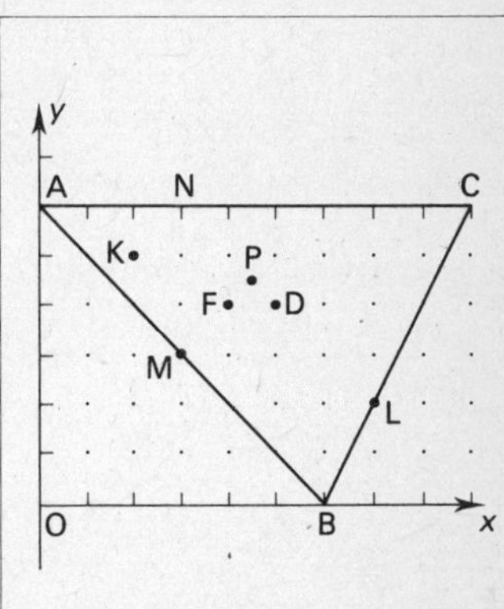

Figure 27

$$\mathbf{OD} = \tfrac{1}{3}\mathbf{OC} + \tfrac{2}{3}\mathbf{OM} = \tfrac{2}{3}(\tfrac{1}{2}\mathbf{OA} + \tfrac{1}{2}\mathbf{OB}) + \tfrac{1}{3}\mathbf{OC} = \tfrac{1}{3}\mathbf{OA} + \tfrac{1}{3}\mathbf{OB} + \tfrac{1}{3}\mathbf{OC}$$

Knowing the ratios, there is no need for coordinates; statements 1 and 3 are correct. D is the centroid (centre of mass or gravity) of the triangle. ANSWER B

Example 9
In Figure 27 $\mathbf{OF}$ in terms of $\mathbf{OA}$, $\mathbf{OB}$ and $\mathbf{OC}$ is equal to

1 $\frac{4}{9}\mathbf{OA} + \frac{2}{9}\mathbf{OB} + \frac{1}{3}\mathbf{OC}$ **2** $\frac{2}{9}\mathbf{OA} + \frac{4}{9}\mathbf{OB} + \frac{1}{3}\mathbf{OC}$
3 $\frac{2}{9}\mathbf{OA} + \frac{1}{3}\mathbf{OB} + \frac{4}{9}\mathbf{OC}$

All three answers are different so only one can be right.
F divides BN in the ratio $2:1 \Rightarrow \mathbf{OF} = \frac{2}{3}\mathbf{ON} + \frac{1}{3}\mathbf{OB}$
N divides AC in the ratio $1:2 \Rightarrow \mathbf{ON} = \frac{1}{3}\mathbf{OA} + \frac{2}{3}\mathbf{OC}$

$$\mathbf{OF} = \tfrac{2}{3}(\tfrac{1}{3}\mathbf{OA} + \tfrac{2}{3}\mathbf{OB}) + \tfrac{1}{3}\mathbf{OB} = \tfrac{2}{9}\mathbf{OA} + \tfrac{1}{3}\mathbf{OB} + \tfrac{4}{9}\mathbf{OC}$$

Notice that the coefficients add up to 1. Statement 3 only is correct. ANSWER E

Example 10

The length of the vector $3\mathbf{i}+4\mathbf{j}+12\mathbf{k}$ is

1 $3+4+12$ **2** 13 **3** $\sqrt{3^2+4^2+12^2}$

To find the length of a vector in three dimensions we use a three-dimensional form of Pythagoras' theorem as in answer 3.

Statements 3 and 2 give the same value 13 and are both correct. As a simple check, remember that a rule for vectors must be true in two dimensions as well as three, and discounting the third component ($12\mathbf{k}$) we know that the length of $3\mathbf{i}+4\mathbf{j}$ is 5 (3,4,5 triangle) so statement 1 ($3+4+12$) cannot possibly be correct.

ANSWER C

Example 11

The angle θ between the vectors $\mathbf{a}$ and $\mathbf{b}$ is given by

1 $\cos\theta = \dfrac{\mathbf{a}\cdot\mathbf{b}}{ab}$ **2** $\tan\theta = \dfrac{|\mathbf{a}+\mathbf{b}|}{ab}$ **3** $\sin\theta = \dfrac{|\mathbf{a}\times\mathbf{b}|}{ab}$

By definition $\mathbf{a}\cdot\mathbf{b} = ab\cos\theta$ where $a=|\mathbf{a}|$ and $b=|\mathbf{b}|$. $\mathbf{a}\cdot\mathbf{b}$ can be calculated from components and θ determined, so statement 1 is correct. Statement 2 is incorrect: if $\mathbf{a}=\mathbf{i}$ and $\mathbf{b}=\mathbf{j}$ then $|\mathbf{a}+\mathbf{b}| = |\mathbf{i}+\mathbf{j}| = \sqrt{2}$. Statement 2 gives $\tan\theta = \dfrac{\sqrt{2}}{1} \Rightarrow \theta = \tan^{-1}\sqrt{2} = 54.7°$, but $\mathbf{i}$ and $\mathbf{j}$ are perpendicular.

The vector product $\mathbf{a}\times\mathbf{b} = ab\sin\theta\,\mathbf{n}$ where $\mathbf{n}$ is a unit vector perpendicular to $\mathbf{a}$ and $\mathbf{b}$. The product $\mathbf{a}\times\mathbf{b}$ can be calculated from components and $\sin\theta$ then calculated. Statement 3 is correct. ANSWER B

These methods for finding the angle between vectors are used in Example 25.

Example 12

The angle between the diagonals of a cube is

1 $90°$ **2** $60°$ **3** $\cos^{-1}\frac{1}{3}$

There are 8 diagonals in a cube, two of which, AOC and BOD, are shown in Figure 28. With O as the origin and $A(1,1,1)$ and $B(1,1,-1)$, $AB = 2$ and $AD = 2\sqrt{2}$.

The angle between the diagonals is either $\angle BOA$ or $\angle AOD$ which are different but add up to $180°$. In $\triangle OAM$

$$\cos\theta = OM/AO = \sqrt{2}/\sqrt{3}$$

$$\cos 2\theta = 2\cos^2\theta = 1$$

$$= 1\tfrac{1}{3} - 1 = \tfrac{1}{3}$$

$$BOA = 2\theta = \cos^{-1}\tfrac{1}{3} = 70.5°$$

Statement 3 only is correct.

ANSWER E

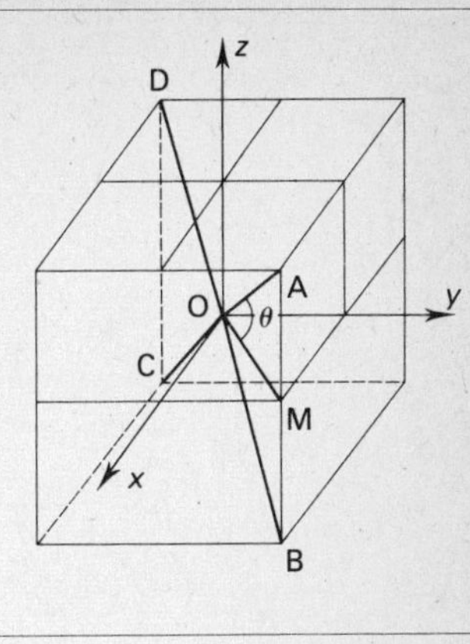

Figure 28

Multiple choice questions (Type C) For each question two statements are given. Answer

A if 1 implies 2 and 2 implies 1
B if 2 implies 1 but 1 does not imply 2
C if 1 implies 2 but 2 does not imply 1
D if 1 denies 2 and 2 denies 1
E if none of these relations holds

Example 13

a and **b** are vectors

1 $\mathbf{a}+\mathbf{b}$ and $\mathbf{a}-\mathbf{b}$ are perpendicular
2 **a** and **b** are the same length

If $\mathbf{a} = \mathbf{OA}$ and $\mathbf{b} = \mathbf{OB}$ then $\mathbf{OC} = \mathbf{a}+\mathbf{b} = \mathbf{OA}+\mathbf{OB}$ and $OACB$ forms a parallelogram with $\mathbf{OC} = \mathbf{a}+\mathbf{b}$ and $\mathbf{BA} = \mathbf{a}-\mathbf{b}$.

If **a** and **b** are the same length, $OACB$ is a rhombus $\Rightarrow$ the diagonals are perpendicular $\Rightarrow (\mathbf{a}+\mathbf{b})\cdot(\mathbf{a}-\mathbf{b}) = 0$

$\Rightarrow a^2 - b^2 = 0 \Rightarrow a^2 = b^2$, i.e., the lengths of **a** and **b** are equal. The argument is reversible so statement 1 implies statement 2 and vice versa.

ANSWER A

153

> **Example 14**
> **1** $\mathbf{AB} + \mathbf{BC} = \mathbf{AC}$ **2** $|\mathbf{AB}| + |\mathbf{BC}| = |\mathbf{AC}|$

$\mathbf{AB} + \mathbf{BC} = \mathbf{AC}$ is always true for any positions of A, B and C, from the addition rule for vectors.

Considering the triangle ABC, the sum of the lengths of any two sides is always greater than the length of the third side, so $|\mathbf{AB}| + |\mathbf{BC}| \geqslant |\mathbf{AC}|$ and the equality only occurs when B lies on the line AC and between A and C.

If statement 2 is true, A, B and C are collinear (in the same straight line) and statement 1 will also be true. Statement 2 implies statement 1 but statement 1 does not imply statement 2.   ANSWER B

> **Example 15**
> **1** $\mathbf{a}$ is perpendicular to $\mathbf{b}$ **2** $\mathbf{a} \cdot \mathbf{b} = 0$

$\mathbf{a}$ and $\mathbf{b}$ perpendicular $\Rightarrow \mathbf{a} \cdot \mathbf{b} = ab \cos 90° = 0$. Statement 1 implies statement 2.
$\mathbf{a} \cdot \mathbf{b} = 0 \Rightarrow ab \cos\theta = 0 \Rightarrow \cos\theta = 0$ or $\mathbf{a} = \mathbf{0}$ or $\mathbf{b} = \mathbf{0}$
Statement 2 does not necessarily imply statement 1. ANSWER C

> **Example 16**
> $\mathbf{a}$ and $\mathbf{b}$ are non-zero vectors
>
> **1** $\mathbf{a} \times \mathbf{b} = \mathbf{0}$ **2** $\mathbf{a}$ is a multiple of $\mathbf{b}$

The vector product, $\mathbf{a} \times \mathbf{b} = ab \sin\theta\mathbf{n}$ where θ is the angle between $\mathbf{a}$ and $\mathbf{b}$ measured positively from $\mathbf{a}$ to $\mathbf{b}$ and $\mathbf{n}$ is the unit vector perpendicular to $\mathbf{a} + \mathbf{b}$ such that $\mathbf{a}$, $\mathbf{b}$ and $\mathbf{n}$ form a right hand set. $\mathbf{a} = t\mathbf{b} \Rightarrow$ the angle between $\mathbf{a}$ and $\mathbf{b}$ is zero.

$\mathbf{a} = t\mathbf{b} \Rightarrow \mathbf{a} \times \mathbf{b} = ab \sin\theta\mathbf{n} \Rightarrow \mathbf{0}$. Statement 2 implies statement 1.
$\mathbf{a} \times \mathbf{b} = \mathbf{0} \Rightarrow \theta = 0$ since $\mathbf{a}$ and $\mathbf{b} = \mathbf{0} \Rightarrow$ the angle between $\mathbf{a}$ and $\mathbf{b}$ is zero $\Rightarrow \mathbf{a}$ and $\mathbf{b}$ are parallel $\Rightarrow \mathbf{a}$ is a multiple of $\mathbf{b}$.

Statement 1 implies statement 2 and statement 2 implies statement 1.   ANSWER A

Multiple choice questions (Type D) Each question consists of a problem followed by four pieces of information. Decide whether the problem can be solved with one of the four pieces of information omitted and answer

A if 1 could be omitted   **D** if 4 could be omitted
B if 2 could be omitted   **E** if none can be omitted
C if 3 could be omitted

Example 17

Find the coordinates of P where $\mathbf{FP} = t\mathbf{FS} + (1-t)\mathbf{FT}$

1 S is $(1,5)$ **2** T is $(9,1)$ **3** $t = \frac{1}{4}$ **4** F is $(2,2)$

The value of t specifies P in relation to S and T. $t = \frac{1}{4} \Rightarrow$ $\mathbf{FP} = \frac{1}{4}\mathbf{FS} + \frac{3}{4}\mathbf{FT}$ and by the ratio theorem P divides ST in the ratio $3:1$ (P is nearer T).

The position of F is immaterial. P is $\frac{3}{4}$ of the way from S to T and as $\mathbf{ST} = 8\mathbf{i} - 4\mathbf{j}$, if O is the origin

$$\mathbf{OP} = \mathbf{OS} + \tfrac{3}{4}\mathbf{ST} = \mathbf{i} + 5\mathbf{j} + \tfrac{3}{4}(8\mathbf{i} - 4\mathbf{j}) = 7\mathbf{i} + 2\mathbf{j}$$

Using coordinates $\quad P = \frac{1}{4}S + \frac{3}{4}T = \frac{1}{4}(1,5) + \frac{3}{4}(9,1)$
$$= \tfrac{1}{4}(1,5) + \tfrac{1}{4}(27,3) = \tfrac{1}{4}(28,8) = (7,2)$$

ANSWER D

Example 18

The Cartesian equation of the plane ABC is $ax + by + cz = d$. Find the values of a, b, c and d.

1 A is $(3,2,1)$ **2** B is $(1,2,3)$ **3** C is $(6,6,1)$
4 AC has length 5

Three points determine a plane so values 1, 2 and 3 are sufficient. Value 4 is not needed.   **ANSWER D**

Value 4 can be deduced from values 1 and 3. The actual equation is found in Example 22.

Example 19

Find the distance of the point D from the plane ABC.
1 A is $(3,2,1)$ **2** B is $(1,2,3)$ **3** C is $(6,6,1)$
4 D is $(3,4,5)$

The plane is determined by 3 points so values A, B and C are needed to fix the position of the plane. D is then needed to find its distance from the plane ABC. All four facts are needed.

ANSWER E

The working of this example is shown in Example 23 on page 158.

Short questions

> **Example 20**
> If D is the centroid of $\triangle ABC$, show that $\mathbf{DA} + \mathbf{DB} + \mathbf{DC} = \mathbf{0}$. Use Figure 27 for reference and check the result using the coordinates to specify the vectors.

Referring to Figure 27, M is the mid-point of $\mathbf{AB} \Rightarrow \mathbf{DM} = \frac{1}{2}\mathbf{DA} + \frac{1}{2}\mathbf{DB}$

$\mathbf{DA} + \mathbf{DB} = 2\mathbf{DM} = \mathbf{CD} = -\mathbf{DC} \Rightarrow \mathbf{DA} + \mathbf{DB} + \mathbf{DC} = \mathbf{0}$

$\mathbf{DA} + \mathbf{DB} + \mathbf{DC} = (-5\mathbf{i} + 2\mathbf{j}) + (\mathbf{i} - 4\mathbf{j}) + (4\mathbf{i} + 2\mathbf{j}) = \mathbf{0}$

> **Example 21**
> Find the centroid of the tetrahedron $A(0,6,0)$, $B(6,0,0)$, $C(9,6,0)$, $V(5,4,8)$.

The centroid G is the intersection of lines joining each vertex to the centroid of the opposite face. The centroid of $\triangle ABC$ has already been calculated in two dimensions in Example 8 and is denoted by $D(5,4,0)$ in Figure 28.

G divides VD in the ratio $3:1$ so $\mathbf{OG} = \frac{1}{4}\mathbf{OV} + \frac{3}{4}\mathbf{OD}$

$\mathbf{OG} = \frac{1}{4}(5\mathbf{i} + 4\mathbf{j} + 8\mathbf{k}) + \frac{3}{4}(5\mathbf{i} + 4\mathbf{j}) = \frac{1}{4}(20\mathbf{i} + 16\mathbf{j} + 8\mathbf{k}) = 5\mathbf{i} + 4\mathbf{j} + 2\mathbf{k}$

As a check, G should divide AE (E is the centroid of BCV) in the ratio $3:1$ so that

$$\begin{aligned}
\mathbf{OG} = \frac{1}{4}\mathbf{OA} + \frac{3}{4}\mathbf{OE} &= \frac{1}{4}(6\mathbf{j}) + \frac{3}{4}\cdot\frac{1}{3}(\mathbf{OB} + \mathbf{OC} + \mathbf{OV}) \\
&= \frac{1}{4}(6\mathbf{j}) + \frac{1}{4}(6\mathbf{i} + 9\mathbf{i} + 6\mathbf{j} + 5\mathbf{i} + 4\mathbf{j} + 8\mathbf{k}) \\
&= \frac{1}{4}(20\mathbf{i} + 16\mathbf{j} + 8\mathbf{k}) = 5\mathbf{i} + 4\mathbf{j} + 2\mathbf{k}
\end{aligned}$$

From this we can see that
$$\begin{aligned}
\mathbf{OG} = \frac{1}{4}(\mathbf{OA} + \mathbf{OB} + \mathbf{OC} + \mathbf{OV}) &= \frac{1}{2}(\frac{1}{2}\mathbf{OA} + \frac{1}{2}\mathbf{OB}) + \frac{1}{2}(\frac{1}{2}\mathbf{OC} + \frac{1}{2}\mathbf{OV}) \\
&= \frac{1}{2}\mathbf{OM} + \frac{1}{2}\mathbf{OW}
\end{aligned}$$

M is the mid-point of AB and W is the mid-point of CV.
G is the centre of the line joining the mid-points of opposite edges.

> **Example 22**
> Find the equation of the plane through $A(3,2,1)$, $B(1,2,3)$, $C(6,6,1)$ and deduce the points X, Y and Z where the plane meets the axes.

A plane has a vector equation and a Cartesian equation.

Method 1 The Cartesian equation is given by $ax + by + cz = d$

$A(3,2,1)$ lies on the plane $\Rightarrow 3a + 2b + c = d$ (1)

$B(1,2,3)$ lies on the plane $\Rightarrow a + 2b + 3c = d$ (2)

$C(6,6,1)$ lies on the plane $\Rightarrow 6a + 6b + c = d$ (3)

$(1) - (2)$ gives $2a - 2c = 0 \Rightarrow a = c$

With $a = c$ (2) becomes $4a + 2b = d$ $5a = 2d \Rightarrow a = c = \frac{2}{5}d$

                 (3) becomes $7a + 6b = d$

$2b = d - 4a = d - \frac{8}{5}d = -\frac{3}{5}d \Rightarrow b = -\frac{3}{10}d$

$ax + by + cz = d$ becomes $\frac{2}{5}dx - \frac{3}{10}dy + \frac{2}{5}dz = d \Rightarrow 4x - 3y + 4z = 10$.

The plane meets the x-axis where $y = 0$ and $z = 0 \Rightarrow x = 2\frac{1}{2}$ and X is $(2\frac{1}{2},0,0)$, Y is $(0, -3\frac{1}{3}, 0)$ and Z is $(0,0,2\frac{1}{2})$.

Method 2 The vector equation of the plane is given by
$\mathbf{r} = x\mathbf{i} + y\mathbf{j} + z\mathbf{k} = \mathbf{OA} + s\mathbf{AB} + t\mathbf{AC}$

$$\Rightarrow \mathbf{r} = \mathbf{OA} + s(\mathbf{OB} - \mathbf{OA}) + t(\mathbf{OC} - \mathbf{OA})$$
$$\Rightarrow \mathbf{r} = 3\mathbf{i} + 2\mathbf{j} + \mathbf{k} + s(-2\mathbf{i} + 2\mathbf{k}) + t(3\mathbf{i} + 4\mathbf{j})$$

Different values of s and t give different points on the plane, e.g.,
$s = t = 0 \Rightarrow \mathbf{r} = \mathbf{OA}$; $s = 1, t = 0 \Rightarrow \mathbf{r} = \mathbf{OB} = \mathbf{i} + 2\mathbf{j} + \mathbf{k}$.

To find the Cartesian equation s and t must be eliminated.
Equating coefficients of

$\mathbf{i} \Rightarrow x = 3 - 2s + 3t$ (1)

$\mathbf{j} \Rightarrow y = 2 + 4t \Rightarrow t = \frac{1}{4}(y - 2)$

$\mathbf{k} \Rightarrow z = 1 + 2s \Rightarrow s = \frac{1}{2}(z - 1)$

Substituting for s and t in $(1) \Rightarrow x = 3 - (z - 1) + \frac{3}{4}(y - 2)$
$$\Rightarrow 4x = 12 - 4z + 4 + 3y - 6$$
$$\Rightarrow 4x - 3y + 4z = 10$$

X, Y and Z are found as in Method 1.

Method 3 For the plane $ax + by + cz = d$ the vector $a\mathbf{i} + b\mathbf{j} + c\mathbf{k}$ is the normal vector, i.e., it is a vector which is perpendicular to the plane and in effect gives the 'direction' of the plane. This normal vector is perpendicular to all vectors in the plane. To find it, form the vector product of two vectors in the plane (in different directions), e.g., $\mathbf{AB}$ and $\mathbf{AC}$

$$\mathbf{AB} \times \mathbf{AC} = (-2\mathbf{i} + 2\mathbf{k}) \times (3\mathbf{i} + 4\mathbf{j}) = \begin{vmatrix} \mathbf{i} & \mathbf{j} & \mathbf{k} \\ -2 & 0 & 2 \\ 3 & 4 & 0 \end{vmatrix} = -8\mathbf{i} + 6\mathbf{j} - 8\mathbf{k}$$

The plane ABC has the form $-8x + 6y - 8z = d$ and since $A(1,2,3)$

lies in the plane $-8+12-24 = d \Rightarrow d = -20$.

The plane has equation $-8x+6y-8z = -20 \Rightarrow 4x-3y+4z = 10$.

Method 3 involves working out 3 second order determinants rather than 3 simultaneous equations and is therefore the quickest method.

Example 23

Find the distance of the point $D(3,4,5)$ from the plane through $A(1,2,3)$, $B(3,2,1)$ and $C(6,6,1)$.

Plane ABC was found in the previous example to be $4x-3y+4z = 10$. The distance of the point (x_1,y_1,z_1) from the plane $ax+by+cz = d$ is given by $\dfrac{ax_1+by_1+cz_1-d}{\sqrt{(a^2+b^2+c^2)}} =$

$\dfrac{12-12+20-10}{\sqrt{(4^2+(-3)^2+4^2)}} = \dfrac{10}{\sqrt{41}} \simeq 1.56$. The distance from the origin O to the plane ABC is $-10/\sqrt{41} \simeq -1.56$, the negative sign indicating that O is on the opposite side of the plane to D.

Long questions

Example 24

With reference to Figure 27, (a) find the coordinates of P, the circumcentre of ABC and check that the length of vectors $\mathbf{PA} = \mathbf{a}$, $\mathbf{PB} = \mathbf{b}$ and $\mathbf{PC} = \mathbf{c}$ are equal.

(b) Defining $\mathbf{PH} = \mathbf{h} = \mathbf{a}+\mathbf{b}+\mathbf{c}$ show that $\mathbf{CH}$ is perpendicular to $\mathbf{AB}$ and $\mathbf{BH}$ is perpendicular to $\mathbf{AC}$. Use this to find H.

(c) How does H relate to $\triangle ABC$ and what is the relation between P, D and H.

(a) The circumcentre of the triangle is the intersection of the mediators (perpendicular bisectors) of the sides.

The mediator of AB is MF whose equation is $y = x$.

The mediator of AC is $x = 4\frac{1}{2}$.

These lines intersect at $P(4\frac{1}{2},4\frac{1}{2})$ so P is the circumcentre.

(b) $\mathbf{PA} = -4\frac{1}{2}\mathbf{i}+1\frac{1}{2}\mathbf{j}$; $\mathbf{PB} = 1\frac{1}{2}\mathbf{i}-4\frac{1}{2}\mathbf{j}$; $\mathbf{PC} = 4\frac{1}{2}\mathbf{i}+1\frac{1}{2}\mathbf{j}$

These three vectors have length $\sqrt{(1\frac{1}{2}^2+4\frac{1}{2}^2)} = \frac{1}{2}\sqrt{(9+81)} = \frac{1}{2}\sqrt{90} = 1\frac{1}{2}\sqrt{10}$. Therefore $\mathbf{a}$, $\mathbf{b}$ and $\mathbf{c}$ have the same length.

$$\mathbf{h} = \mathbf{a}+\mathbf{b}+\mathbf{c} \Rightarrow \mathbf{h}-\mathbf{c} = \mathbf{a}+\mathbf{b}$$

which is perpendicular to $\mathbf{a}-\mathbf{b}$ since $\mathbf{a}$ and $\mathbf{b}$ are equal in length.

The parallelogram $OACB$ where $\mathbf{OA}=\mathbf{a}$, $\mathbf{OB}=\mathbf{b}$, $\mathbf{OC}=\mathbf{a}+\mathbf{b}$ has diagonals OC and AB represented by $\mathbf{OC}=\mathbf{a}+\mathbf{b}$ and $\mathbf{BA}=\mathbf{a}-\mathbf{b}$. If $\mathbf{a}$ and $\mathbf{b}$ are equal in length $OACB$ is a rhombus and the diagonals are perpendicular.

$\mathbf{a}$ and $\mathbf{b}$ have the same length $\Leftrightarrow \mathbf{a}+\mathbf{b}$ and $\mathbf{a}-\mathbf{b}$ are perpendicular
$\mathbf{h}-\mathbf{c}=\mathbf{BH}$ and $\mathbf{a}-\mathbf{b}=\mathbf{BA}\Rightarrow CH$ and AB are perpendicular.
Similarly $\mathbf{h}-\mathbf{b}=\mathbf{a}+\mathbf{c}$ which is perpendicular to $\mathbf{a}-\mathbf{c}$.
$\mathbf{h}-\mathbf{b}=\mathbf{BH}$ and $\mathbf{a}-\mathbf{c}=\mathbf{CA}\Rightarrow BH$ and AC are perpendicular.
CH and BH are altitudes of the $\triangle ABC$ which intersect at H with coordinates $(6,3)$.
As a check $\mathbf{h}=\mathbf{a}+\mathbf{b}+\mathbf{c}=\mathbf{PA}+\mathbf{PB}+\mathbf{PC}=1\tfrac{1}{2}\mathbf{i}-1\tfrac{1}{2}\mathbf{j}=\mathbf{PH}$
$\mathbf{OH}=\mathbf{OP}+\mathbf{PH}=4\tfrac{1}{2}\mathbf{i}+4\tfrac{1}{2}\mathbf{j}+1\tfrac{1}{2}\mathbf{i}-1\tfrac{1}{2}\mathbf{j}=6\mathbf{i}+3\mathbf{j}$ as above.

(c) H is the orthocentre of $\triangle ABC$ (intersection of the altitudes). By a similar argument to those above, HA and BC are perpendicular, P(circumcentre), D(centroid) and H(orthocentre) are collinear and D divides PH in the ratio $1:2$.

Example 25
Find the angle between the vectors $\mathbf{a}=\mathbf{i}+2\mathbf{j}+3\mathbf{k}$ and $\mathbf{b}=3\mathbf{i}-2\mathbf{j}+\mathbf{k}$ by using (i) the scalar product and (ii) the vector product.

(i) $\mathbf{a}\cdot\mathbf{b}=(\mathbf{i}+2\mathbf{j}+3\mathbf{k})\cdot(2\mathbf{i}+2\mathbf{j}+\mathbf{k})=3-4+3=2$
$\mathbf{a}\cdot\mathbf{b}=ab\cos\theta$ where $a=|\mathbf{a}|=\sqrt{1^2+2^2+3^2}=\sqrt{14}$ and
$b=|\mathbf{b}|=\sqrt{14}\Rightarrow\sqrt{14}\sqrt{14}\cos\theta=2\Rightarrow\cos\theta=\tfrac{1}{7}\Rightarrow\theta=81.8°$

(ii)
$$\mathbf{a}\times\mathbf{b}=\begin{vmatrix}\mathbf{i}&\mathbf{j}&\mathbf{k}\\1&2&3\\3&-2&1\end{vmatrix}=8\mathbf{i}+8\mathbf{j}-8\mathbf{k}\Rightarrow|\mathbf{a}\times\mathbf{b}|$$
$$=\sqrt{8^2+8^2+(-8)^2}=8\sqrt{3}$$

$|\mathbf{a}\times\mathbf{b}|=ab\sin\theta$ where $a=b=\sqrt{14}\Rightarrow\sqrt{14}\sqrt{14}\sin\theta=8\sqrt{3}$
$$\Rightarrow\sin\theta=4\sqrt{\tfrac{3}{7}}\Rightarrow\theta=81.8°$$

As a check $\sin^2\theta+\cos^2\theta=\dfrac{16\times3}{49}+\dfrac{1}{49}=\dfrac{49}{49}=1.$

Example 26
Find the line of intersection of the planes $x+2y+3z=14$ and $2x+5y-z=9$ and find where this line intersects the plane $3x+8y-4z=7$.

The equations $\dfrac{x-1}{3} = \dfrac{y-2}{4} = \dfrac{z-3}{5} = t$ or $x = 1+3t,\ y = 2+4t,$ $z = 3+5t$ represent a straight line through the point $(1,2,3)$ in the direction $3\mathbf{i}+4\mathbf{j}+5\mathbf{k}$.

Solving the equations $x+2y+3z = 14 \hspace{4cm} (1)$
$$2x+5y-z = 9 \hspace{4cm} (2)$$
in that form
$(1)+3\times(2) \Rightarrow 7x+17y = 41$ or $7x-41 = -17y$
$(2)-2\times(1) \Rightarrow y-7z = -19$ or $y = 7z-19$

The line of intersection is

$$\dfrac{x-\frac{41}{7}}{-17} = \dfrac{y}{7} = \dfrac{z-\frac{19}{7}}{1} = t \quad \text{or} \quad \left.\begin{array}{l} x = -17t+\frac{41}{7} \\[4pt] y = 7t \\[4pt] z = t+\frac{19}{7} \end{array}\right\} \hspace{2cm} A$$

To find where this line meets the plane $3x+8y-4z = 7$ we find the value of t which allows equations A to satisfy $3x+8y-4z = 7$.
$\Rightarrow 3(-17t+\frac{41}{7})+56t-4(t+\frac{19}{7}) = 7$
$\Rightarrow -51t+52t = 7+\frac{76}{7}-\frac{123}{7} \Rightarrow t = \dfrac{49+76-123}{7} = \frac{2}{7}$
$t = \frac{2}{7} \Rightarrow x = \frac{41}{7}-\frac{34}{7} = 1;\ y = 2;\ z = \frac{21}{7} = 3$
The three planes intersect in the single point $(1,2,3)$.

If the line of intersection of the first two planes turns out to be parallel to the third plane, there will be no solution to the equation. This will be true if the direction vector of the line of intersection is perpendicular to the normal vector of the third plane. If the line of intersection lies completely in the third plane, the t equation will be satisfied by all values of t.

Example 27
In Figure 27 BF meets AC at N. Find X, the point where CF meets AB, and Y, the point where AF meets BC. Check Ceva's theorem, i.e., if CX, AY and BN meet at the same point, F, then $\dfrac{AX}{XB} \times \dfrac{BY}{YC} \times \dfrac{CN}{NA} = +1$.

In two dimensions it is easy to work with coordinates and use the Cartesian equations of each line.
From the diagram, N is $(3,6)$ and $AN:NC = 1:2$.
The equation of BC is $y = 2(x-6) = 2x-12$ and the equation of AF is $y = 6-\frac{1}{2}x$.

These lines intersect at Y where $2x - 12 = 6 - \frac{1}{2}x \Rightarrow 2\frac{1}{2}x = 18$

$$\Rightarrow x = \tfrac{36}{5} = 7.2 \text{ and } y = 14.4 - 12 = 2.4$$

$BY:YC = 2.4:(6-2.4) = 2.4:3.6 = 2:3$.
The equation of **AB** is $x+y = 6$ or $y = 6-x$.
The equation of **CF** is $y - 4 = \frac{2}{5}(x-4) = \frac{2}{5}x - \frac{8}{5}$ or $y = 0.4x + 2$.
These lines intersect at X, where $6 - x = 0.4x + 2.4$.

$$\Rightarrow 1.4x = 3.6 \Rightarrow 7x = 18 \Rightarrow x = \tfrac{18}{7} \Rightarrow y = 6 - 2\tfrac{4}{7} = 3\tfrac{3}{7} = \tfrac{24}{7}$$

$$X = (2\tfrac{4}{7}, 3\tfrac{3}{7}) \Rightarrow AX:XB = 2\tfrac{4}{7}:(6 - 2\tfrac{4}{7}) = 2\tfrac{4}{7}:3\tfrac{3}{7} = 18:24 = 3:4$$

$$\frac{AX}{XB} \times \frac{BY}{YC} \times \frac{CN}{NA} = \tfrac{3}{4} \times \tfrac{2}{3} \times \tfrac{2}{1} = 1$$

In three dimensions it may be easier to work with vectors and use the vector equations of lines. This method is as follows.

The vector equation of BC is $\mathbf{r} = \begin{pmatrix} x \\ y \end{pmatrix} = x\mathbf{i} + y\mathbf{j} = \mathbf{OB} + t\mathbf{BL}$

$$\Rightarrow \mathbf{r} = \begin{pmatrix} 6 \\ 0 \end{pmatrix} + t\begin{pmatrix} 1 \\ 2 \end{pmatrix} = \begin{pmatrix} 6+t \\ 2t \end{pmatrix} \text{ or } (6+t)\mathbf{i} + 2t\mathbf{j}$$

The vector equation of AF is $\mathbf{r} = \mathbf{OA} + s\mathbf{AK} = 6\mathbf{j} + s(2\mathbf{i} - \mathbf{j}) = 2s\mathbf{i} + (6-s)\mathbf{j}$. These intersect at Y where $(6+t)\mathbf{i} + 2t\mathbf{j} = 2s\mathbf{i} + (6-s)\mathbf{j}$

$$\Rightarrow 6 + t = 2s,\ 2t = 6 - s \Rightarrow 6 + t = 2(6 - 2t) \Rightarrow (5t = 6) \Rightarrow t = 1.2$$
$$\Rightarrow s = 6 - 2t = 6 - 2.4 = 3.6$$

$t = 1.2$ and $s = 3.6$ both give Y as $(7.2, 2.4)$ but for this question there is no need to find Y.
$t = 1.2 \Rightarrow \mathbf{BY} = 1.2(\mathbf{i} + 2\mathbf{j})$ and
$\mathbf{BC} = 3(\mathbf{i} + 2\mathbf{j}) \Rightarrow \mathbf{YC} = (3 - 1.2)(\mathbf{i} + 2\mathbf{j})$
So $BY:YC = 1.2:(3 - 1.2) = 1.2:1.8 = 6:9 = 2:3$
AB has equation $\mathbf{r} = \mathbf{OA} + t(\mathbf{i} - \mathbf{j}) = 6\mathbf{j} + t(\mathbf{i} - \mathbf{j}) = t\mathbf{i} + (6 - t)\mathbf{j}$
FC has equation $\mathbf{r} = \mathbf{OF} + s(5\mathbf{i} + 2\mathbf{j}) = 4\mathbf{i} + 4\mathbf{j} + s(5\mathbf{i} + 2\mathbf{j})$
$$= (4 + 5s)\mathbf{i} + (4 + 2s)\mathbf{j}$$
At X, $t = 4 + 5s$ and $6 - t = 4 + 2s \Rightarrow s = -\tfrac{2}{7} \Rightarrow t = 4 - \tfrac{10}{7} = 2\tfrac{4}{7}$.
$t = 2\tfrac{4}{7} \Rightarrow AX:XB = 2\tfrac{4}{7}:(6 - 2\tfrac{4}{7}) = 2\tfrac{4}{7}:3\tfrac{3}{7} = 18:24 = 3:4$, giving the same result.

If you are unfamiliar with vector methods, Method 2 seems longer and more complicated. Some books and examination boards prefer the matrix notation (brackets) and others prefer **i**s and **j**s as they are easier to print and more economical on paper. The best notation largely depends on the notation you are brought up on or use most. The matrix notation automatically separates **i**s (top component) from **j**s (bottom or middle in 3 dimensions) and **k**s (bottom component), whereas it is easy to confuse these in the **i**, **j** and **k** notation. Perhaps it is easier to change from vector matrix form to Cartesian form when in three dimensions. Unfortunately (as in functional notation) mathematicians are saddled with several notations and have to be conversant with all of them.

> ### Example 28
> Find the shortest distance between the lines $x = 1+3t$, $y = 2+4t$, $z = 3+5t$ and $x = 5-s$, $y = -2+8s$, $z = 1-7s$.

The first equation represents a line through A in the direction $(3\mathbf{i}+4\mathbf{j}+5)$. The second represents a line through B in the direction $-\mathbf{i}+8\mathbf{j}-7\mathbf{k}$. (see Figure 29).

The shortest distance between two skew lines is CD, which is at right angles to both lines.

Consider $\dfrac{\mathbf{BA} \times \mathbf{q}}{q}$ which repre-

sents a vector of magnitude $BA \sin \theta \, (= AD)$ perpendicular to BA and BD, i.e., perpendicular to BD and DA, i.e., in direction **n**.

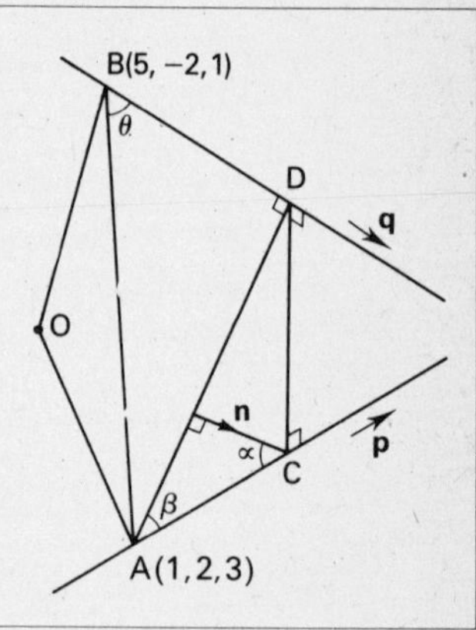

Figure 29

$$\left(\frac{\mathbf{BA} \times \mathbf{q}}{q}\right) \cdot \frac{\mathbf{p}}{p} = AD \cos \alpha \quad \text{where } \alpha \text{ is the angle between } \mathbf{n} \text{ and } \mathbf{p}.$$

$$AD \cos \alpha = AD \sin \beta = CD = \frac{(\mathbf{BA} \times \mathbf{q}) \cdot \mathbf{p}}{pq} = \frac{\mathbf{p} \cdot \mathbf{BA} \times \mathbf{q}}{pq}$$

$$= (3\mathbf{i}+4\mathbf{j}+5\mathbf{k}) \cdot \begin{vmatrix} \mathbf{i} & \mathbf{j} & \mathbf{k} \\ -4 & 4 & 2 \\ -1 & 8 & -7 \end{vmatrix} \div pq$$

$$pq \cdot CD = \begin{vmatrix} 3 & 4 & 5 \\ -4 & 4 & 2 \\ -1 & 8 & -7 \end{vmatrix} = \begin{vmatrix} 3 & 4 & 5 \\ -1 & 8 & 7 \\ -1 & 8 & -7 \end{vmatrix} = \begin{vmatrix} 3 & 4 & 5 \\ -1 & 8 & 7 \\ 0 & 0 & 14 \end{vmatrix}$$

$$= 14 \times 28 = +392$$

$$CD = \frac{392}{pq} = \frac{392}{\sqrt{114}\sqrt{50}} = \frac{392}{\sqrt{5700}} = \frac{392}{75.5} = 5.19$$

Chapter 12

Groups and Algebraic Structure

Multiple choice questions (Type A) Select the correct answer

Example 1

For the set of integers which of the following operations is not closed?

A addition   **B** subtraction   **C** multiplication   **D** division
E taking the larger number

For the set of integers (whole numbers, both positive and negative, and zero) take two numbers and see whether the result of each operation produces an integer.

Answers **A**, **B** and **C** will always give another integer, e.g., $2 + -1 = 1; 2 - -3 = 5; 2 \times -3 = -6$. Taking the larger number (**E**) will always give one of the numbers you started with, i.e., an integer.

However $2 \div -3 = -\frac{2}{3}$, which is a fraction and belongs to the set of rational numbers. Therefore answer D is not closed.

ANSWER D

Example 2

Which of the following sets of integers does not form a group under addition?

A integers reduced modulo 4   **B** modulo 5   **C** modulo 6
D modulo 7   **E** they all form groups

Three of the combination tables are shown in Figure 30. Each is closed, associative, has identity 0 and inverses for every element. The integers form a group under addition for all moduli.

ANSWER E

modulo 4				
+	0	1	2	3
0	0	1	2	3
1	1	2	3	0
2	2	3	0	1
3	3	0	1	2

modulo 5					
+	0	1	2	3	4
0	0	1	2	3	4
1	1	2	3	4	0
2	2	3	4	0	1
3	3	4	0	1	2
4	4	0	1	2	3

modulo 6						
+	0	1	2	3	4	5
0	0	1	2	3	4	5
1	1	2	3	4	5	0
2	2	3	4	5	0	1
3	3	4	5	0	1	2
4	4	5	0	1	2	3
5	5	0	1	2	3	4

Figure 30

Example 3

For which of the following values do the integers, reduced modulo p, **not** form a group under multiplication modulo p?

A $p = 3$ **B** $p = 5$ **C** $p = 7$ **D** $p = 9$ **E** $p = 11$

A group is formed for each value as long as p is prime, so answer D is the only value which does not form a group. The reason is that in D we have divisors of zero. This means that there are elements which multiply together to form zero, e.g., $3 \times 3 = 0$, $3 \times 6 = 0$. The set is not closed, zero being excluded. Even if zero were included, 0 would not have an inverse. The identity is 1 and there is no element which combines with 0 to produce 1 so 0 has no inverse. This does not occur when p is prime since in that case p can have no factors.   ANSWER D

Example 4

Which of the following systems is not distributive?

A multiplication over addition for real numbers
B multiplication over subtraction for real numbers
C addition over multiplication for real numbers
D union over intersection for sets
E intersection over union for sets

The operation $*$ is distributive over $@$ if $a * (b @ c) = (a * b) @ (a * c)$.

Answers A and B are true, e.g., $5 \times (4 \pm 3) = (5 \times 4) \pm (5 \times 3)$.
$5 + (3 \times 4) = 17$ but $(5 + 3) \times (5 + 4) = 72$, so addition is not distributive over multiplication.

Answers D and E are both true. For example, $X \cup (Y \cap Z) = (X \cup Y) \cap (X \cup Z)$. This is best demonstrated by shading the respective sets on a Venn diagram. **ANSWER C**

Example 5

Which of the following equations has no solution if $x \in Z_5$, the set of integers modulo 5?

A $2x = 3$ **B** $2x + 3 = 0$ **C** $2(x + 3) = 0$ **D** $x^2 + 1 = 0$
E $x^2 + 2 = 0$

A $2x = 3 \Rightarrow x = 4$ since $2 \times 4 = 8 = 3 \pmod 5$
B $2x + 3 = 0 \Rightarrow 2x = 2 \Rightarrow x = 1$
C $2(x + 3) = 0 \Rightarrow x + 3 = 0 \Rightarrow x = 2$
D $x^2 + 1 = 0 \Rightarrow x^2 = 4 \Rightarrow x = 2$ or 3 since $3^2 = 4$
E $x^2 + 2 = 0 \Rightarrow x^2 = 3 \Rightarrow$ no solution

$1^2 = 1, 2^2 = 4, 3^2 = 4, 4^2 = 1$, so $x^2 = 3$ has no solution. **ANSWER E**

In D, $x^2 + 1 = x^2 + 5x + 6 = (x + 2)(x + 3) = 0$
$$x + 2 = 0 \Rightarrow x = 3$$
$$x + 3 = 0 \Rightarrow x = 2$$

Multiple choice questions (Type B) Answer according to the table

A	B	C	D	E
1, 2, 3	1, 3	2, 3	2	3
correct	only	only	only	only

Example 6

The set of matrices of the form $\begin{pmatrix} a & -b \\ b & a \end{pmatrix}$ under the operation of matrix multiplication

1 is closed. **2** is commutative. **3** is associative.

To examine the property of closure consider two matrices of this type, i.e.
$$A \quad \times \quad C \quad = \quad AC$$
$$\begin{pmatrix} a & -b \\ b & a \end{pmatrix} \begin{pmatrix} c & -d \\ d & c \end{pmatrix} = \begin{pmatrix} ac - bd & -ad - bc \\ bc + ad & -bd + ac \end{pmatrix}$$

Since in the product AC the components on the leading diagonal are equal and those on the reverse diagonal are equal and opposite in sign, the product matrix has the same form as A and C. Make sure that the components of each matrix a, b, c and d belong to the same set of numbers. For instance, the question may have specified that they be Z(integers), or Q(rationals), or R(real numbers) in which case the elements of the product AC must belong to these sets also. In fact, in each of these cases this is true. The set is closed.

To investigate commutativity evaluate CA

$$C \quad \times \quad A \quad = \quad CA$$

$$\begin{pmatrix} c & -d \\ d & c \end{pmatrix}\begin{pmatrix} a & -b \\ b & a \end{pmatrix} \quad \begin{pmatrix} ac-bd & -bc-da \\ ad+bc & -db+ac \end{pmatrix} = AC$$

$AC = CA \Rightarrow$ commutativity (which is not true in general for matrix multiplication), so the operation is commutative.

Matrix multiplication is associative for all compatible matrices. In this case for $E = \begin{pmatrix} e & -f \\ f & e \end{pmatrix}$ evaluate $(AC)E$ and $A(CE)$, the brackets indicating which pair to work out first.
Statements 1, 2 and 3 are all correct.   ANSWER A

This question can also be answered with a little knowledge of transformations. If the determinant of these matrices were equal to 1, each matrix would represent a rotation in two dimensions about the origin. In that case the properties of closure, commutativity and associativity are easily verified. As it is, matrix A represents a rotation (of angle $\arctan b/a$) combined with an enlargement of area scale factor $\triangle = a^2+b^2$. This combination of transformations is called a spiral similarity. The order of performing the rotation and enlargement does not affect the result. Knowing that rotations and enlargements are closed, associative and commutative could easily give the results above.

Example 7
Which of the following combination tables represents a group structure?

1

*	a	b	c	d
a	a	b	c	d
b	d	a	b	c
c	c	d	a	b
d	b	c	d	a

2

*	a	b	c
a	a	b	c
b	b	c	a
c	c	a	b

3

*	a	b	c
a	a	b	c
b	c	a	b
c	b	c	a

Combination table 2 is the cyclic group of order 3. It is isomorphic (same structure) to the integers mod 3 under addition, and also to rotations of 120°, 240° and 360° and, as such, forms a group. When the operation is unspecified it is easier to prove a group by establishing an isomorphic structure, and quoting the result for the isomorphic structure.

Combination tables 1 and 3 have no proper identity. For both tables a is a left-side identity, i.e., $a*b=b$, $a*c=c$, $a*d=d$ in table 1, but no right-side identity exists. In both these cases the operation is not associative.

For 1, $b(cd) = bb = a$ but $(bc)d = bd = c$
For 3, $b(ac) = bc = b$ but $(ba)c = cc = a$ **ANSWER D**

Example 8

For $a, b \in R$ (real numbers) $a*b = \frac{1}{2}(a+b)$. The operation

1 is associative.   **2** is commutative.   **3** has an identity.

$b*a = \frac{1}{2}(b+a) = \frac{1}{2}(a+b)$, so the operation is commutative and statement 2 is true.

$(a*b)*c = \frac{1}{2}[\frac{1}{2}(a+b)+c] = \frac{1}{4}a + \frac{1}{4}b + \frac{1}{2}c$ These are only the
$a*(b*c) = \frac{1}{2}[a+\frac{1}{2}(b+c)] = \frac{1}{2}a + \frac{1}{4}b + \frac{1}{4}c$ same when $a = c$.

Therefore statement 1 is untrue.
If b is the identity element $\frac{1}{2}(a+b) = a \Rightarrow \frac{1}{2}b = \frac{1}{2}a$ or $a = b$. In general the only element which combines with a to produce a is a itself, but an identity has to be the same for every element of the set. Therefore there is no identity element and so statement 3 is false.   **ANSWER D**

Example 9

The elements A and B generate a group where $A^4 = I$ (identity), $A^2 = B^2$ and $AB = BA^3$

1 $AB = BA$ **2** $A^2B = AB^2$ **3** $A^3B = BA$

$AB = BA^3 \Rightarrow AB^2 = BA^3B = BA^2AB = BA^2BA^3 = BAABA^3$

$$= BABA^3A^3 = BABA^2 = BBA^3A^2 = B^2A$$

      Statement 2 is true

This is achieved by changing each AB into BA^3

$$A^3B = AA^2B = B^2BA^3 = BB^2AA^2 = BAB^2A^2 = BAA^2A^2 = BA$$

Statement 3 is true and statements 2 and 3 are true. ANSWER C

Multiple choice questions (Type C) For each question two statements are given. Answer

A if 1 implies 2 and 2 implies 1
B if 2 implies 1 but 1 does not imply 2
C if 1 implies 2 but 2 does not imply 1
D if 1 denies 2 and 2 denies 1
E if none of these relations holds

Example 10

For the matrix $M = \begin{pmatrix} a & -b \\ b & a \end{pmatrix}$, $a,b \in R$

1 M represents a rotation   **2** $a^2 + b^2 = 1$

The matrix representing a rotation of $A°$ about the origin is $\begin{pmatrix} \cos A & -\sin A \\ \sin A & \cos A \end{pmatrix}$. M is of this form if $a = \cos A$ and $b = \sin A$. If $a^2 + b^2 = 1$ then a and b can represent the sine and cosine of an angle, so statement 2 implies statement 1.

If M represents a rotation, the determinant of the matrix must be 1 (no enlargement) so $a^2 + b^2 = 1$. Therefore statement 1 implies statement 2.   ANSWER A

Example 11

For the set of elements $S = (a,b,c,d)$

1 S is a group   **2** Its combination table is a Latin square

---

A Latin square has each row and column consisting of a permutation of the constituent elements. If S is a group and one element is repeated it will mean that either (i) $ab = ac$ if a row has 2 entries the same or (ii) $ba = ca$ if a column has 2 entries the same.
(i) $ab = ac \Rightarrow a^{-1}(ab) = a^{-1}(ac) \Rightarrow (a^{-1}a)b = (a^{-1}a)c \Rightarrow b = c$ and the column headed by c can be deleted.
(ii) $ba = ca \Rightarrow baa^{-1} = caa^{-1} \Rightarrow b = c$ and the c row is deleted.

Notice that we need the associative property and the inverse of a,

a^{-1}. A group must have a Latin square for its combination table. Statement 1 implies statement 2 but the converse is not true. A counter-example is provided in Example 7.   ANSWER C

Example 12
For the set $S = (a,b,c,d)$ under the operation $*$

1 $*$ is associative   **2** $a*(b*c) = (a*b)*c$

The relation in statement 2 must hold for all members of the set S for the operation to be associative. It is not enough for it to hold for only one triple. Statement 2 does not imply statement 1 but statement 1 does imply statement 2.   ANSWER C

Example 13
a and b are elements of a group G

1 $ab = ba$ **2** $a^{-1}b = ba^{-1}$

$ab = ba \Rightarrow a^{-1}ab = a^{-1}ba \Rightarrow b = a^{-1}ba \Rightarrow ba^{-1} = a^{-1}baa^{-1} = a^{-1}b$
$a^{-1}b = ba^{-1} \Rightarrow aa^{-1}b = aba^{-1} \Rightarrow b = aba^{-1} \Rightarrow ba = aba^{-1}a = ab$
Statement 1 implies statement 2 and vice versa.   ANSWER A

Multiple choice questions (Type D) Each question consists of a problem followed by four pieces of information. Decide whether the problem can be solved with one of the four pieces of information omitted and answer

A if 1 could be omitted   **D** if 4 could be omitted
B if 2 could be omitted   **E** if none can be omitted
C if 3 could be omitted

Example 14
The set G under the operation $*$ forms a group.

1 G is closed under $*$ **2** The operation $*$ is associative
3 G contains an identity element
4 Each member of G has an inverse

These are the four properties that define a group. All four are necessary.   ANSWER E

Example 15

Can a group of order greater than 2 be formed from the following matrices under matrix multiplication?

$$\begin{array}{cccc} A & B & C & D \\ \mathbf{1}\begin{pmatrix} -1 & 0 \\ 0 & -1 \end{pmatrix} & \mathbf{2}\begin{pmatrix} 0 & -1 \\ 1 & 0 \end{pmatrix} & \mathbf{3}\begin{pmatrix} 0 & 1 \\ -1 & 0 \end{pmatrix} & \mathbf{4}\begin{pmatrix} 1 & 0 \\ 0 & 1 \end{pmatrix} \end{array}$$

If you know that D is the identity, A represents a half-turn (rotation of 180° about the origin), B is a rotation of $+90°$ and C is a rotation of $+270°(-90°)$ then this question is easy. AB will represent a rotation of 270°, i.e., C, so the set A,B,D will not be closed. This will apply to any combination of 3 of the matrices. For any three matrices taken, there will be a combination which produces the fourth and the four A, B, C, D do form a group.

Without transformations, form the combinations table by performing the matrix multiplications.

The set is closed (from the table), associative (matrix multiplication is associative), the identity is D and all four have inverses.

×	A	B	C	D
A	D	C	B	A
B	C	A	D	B
C	B	D	A	C
D	A	B	C	D

Figure 31

D and A are self-inverse
B is the inverse of C
C is the inverse of B
B and C form an inverse pair

ANSWER E

Short questions

Example 16

Show that the set of non-zero real numbers forms a group under the operation $a * b$ where $a * b = 2ab$.

$a,b \in R$ (real numbers) $\Rightarrow 2ab \in R$ so the set is closed.

$$(a * b) * c = (2ab) * c = 4abc \qquad a * (b * c) = a * (2bc) = 4abc$$

The operation is associative.

If b is the identity element, $a * b = a \Rightarrow 2ab = a \Rightarrow b = \frac{1}{2}$.
$b * a = \frac{1}{2} * a = 2 \cdot \frac{1}{2}a = a$, so $a * \frac{1}{2} = \frac{1}{2} * a = a$ and $\frac{1}{2}$ is the identity.
If b is the inverse of a, $a * b = \frac{1}{2} \Rightarrow 2ab = \frac{1}{2} \Rightarrow b = \frac{1}{4}a$. Each element a
has an inverse $\frac{1}{4}a$, and the set forms a group.

Example 17

Find the group generated by the matrix $M = \begin{pmatrix} 0 & 1 & 0 \\ 0 & 0 & 1 \\ 1 & 0 & 0 \end{pmatrix}$

$$M^2 = \begin{pmatrix} 0 & 1 & 0 \\ 0 & 0 & 1 \\ 1 & 0 & 0 \end{pmatrix} \begin{pmatrix} 0 & 1 & 0 \\ 0 & 0 & 1 \\ 1 & 0 & 0 \end{pmatrix} = \begin{pmatrix} 0 & 0 & 1 \\ 1 & 0 & 0 \\ 0 & 1 & 0 \end{pmatrix}$$

$$M^3 = \begin{pmatrix} 0 & 0 & 1 \\ 1 & 0 & 0 \\ 0 & 1 & 0 \end{pmatrix} \begin{pmatrix} 0 & 1 & 0 \\ 0 & 0 & 1 \\ 1 & 0 & 0 \end{pmatrix} = \begin{pmatrix} 1 & 0 & 0 \\ 0 & 1 & 0 \\ 0 & 0 & 1 \end{pmatrix} = I$$

The matrix M generates the cyclic group of order 3.
This method may be tedious if the group has order 6 or more.
It may be easier to consider the transformation that M represents.
M takes $I(1,0,0)$ to $K(0,0,1)$; $J(0,1,0)$ to $I(1,0,0)$ and $K(0,0,1)$ to
$J(0,1,0)$.
This rotates the triangle IJK to JKI, i.e., a rotation of $120°$. M^2 will
represent a rotation of $240°$ and $M^3 = I$ (Rotation of $360°$). The
rotations form a group of order 3 and so will the matrices.

Example 18
Find the group with the smallest order which contains the
complex number i where elements are complex numbers
under the operation of multiplication.

The smallest group will be that generated by i; in other words all
the elements will be powers of i.
$$i^2 = -1; \quad i^3 = -i; \quad i^4 = 1$$
So the group consists of $(1, i, -1, -i)$.

The set is closed and associative; 1 is the identity; 1 and -1 are
self-inverse and i and $-i$ form an inverse pair.

As an extra question, show that this group is cyclic.

Example 19

Find the order of each element of the group of integers under multiplication modulo 7 and use this to show that the group is cyclic.

The order of an element is the value of the lowest power which gives the identity 1.

$2^2 = 4, 2^3 = 1$, so the order of 2 is 3
$3^2 = 2, 3^3 = 6, 3^4 = 4, 3^5 = 5, 3^6 = 1$ and the order of 3 is 6
$4^2 = 2, 4^3 = 1$; the order of 4 is 3
$5^2 = 4, 5^3 = 6, 5^4 = 2, 5^5 = 3, 5^6 = 1$; the order of 5 is 6
$6^2 = 1$; the order of 6 is 2

Each element generates a subgroup. 3 and 5 generate the complete group so if the elements were written in the order of the powers of either 3 or 5, the group would be seen to be cyclic (see Figure 32).

MULTIPLICATION MODULO 7

$\times$	1	2	3	4	5	6
1	1	2	3	4	5	6
2	2	4	6	1	3	5
3	3	6	2	5	1	4
4	4	1	5	2	6	3
5	5	3	1	6	4	2
6	6	5	4	3	2	1

$\times$	1	3	3^2	3^3	3^4	3^5
1	1	3	2	6	4	5
3	3	2	6	4	5	1
$3^2 = 2$	2	6	4	5	1	3
$3^3 = 6$	6	4	5	1	3	2
$3^4 = 4$	4	5	1	3	2	6
$3^5 = 5$	5	1	3	2	6	4

Figure 32

Example 20

Show that the permutations of three letters (A,B,C) form a group.

There are $3 \times 2 \times 1 = 6$ ways of arranging ABC so there are 6 permutations of the three letters. When performed on the order ABC we get

$$ABC \quad ACB \quad BAC \quad BCA \quad CAB \quad CBA$$

These correspond to transforming the equilateral triangle as in Example 23 (Figure 34). The permutations are in one-one correspondence with the transformations and so they form a group. The permutations corresponding to the rotations BCA and CAB together with the identity ABC are **even** permutations, and ACB, BAC, CBA, corresponding to the opposite isometries, (reflexions) are **odd** permutations.

If you can show that a system is isomorphic to another system then the properties of each carry over.

Example 21
Examine the binary operation $a*b = a+b+1$ for group properties.

Closure depends on which set of numbers a and b are taken from. For N, Z, R, Q and C the set is closed but for the set of even numbers it is not closed.

$$\left. \begin{array}{l} (a*b)*c = (a+b+1)*c = a+b+1+c+1 \\ a*(b*c) = a*(b+c+1) = a+b+c+1+1 \end{array} \right\} \quad \begin{array}{l} \text{The operation} \\ \text{is associative.} \end{array}$$

If b is the identity, $a*b = a \Rightarrow a+b+1 = a \Rightarrow b = -1$
If b is the inverse of a, $a*b = -1 \Rightarrow a+b+1 = -1 \Rightarrow b = -2-a$
The inverse of a is $-2-a$; $(-2-a)*a = -2-a+a+1 = -1$ also.
The sets Z, Q, R and C form groups for the operation.
N does not, as there is no identity element, nor are there any inverses.

Long questions

Example 22
Show that each of the sets $A = (1,3,7,9)$ and $B = (2,4,6,8)$ form groups under multiplication modulo 10. Why is 5 excluded from both sets? Show that A and B are isomorphic.

The combination tables are shown in Figure 33(a) and (b).

The order 1,3,9,7 emphasizes that A is a cyclic group in which the order 1397 appears along each row but starting each time with the heading element. From each table both sets are closed.

MULTIPLICATION MODULO 10

×	1	3	9	7
1	1	3	9	7
3	3	9	7	1
9	9	7	1	3
7	7	1	3	9

(a)

×	2	4	6	8
2	4	8	2	6
4	8	6	4	2
6	2	4	6	8
8	6	2	8	4

(b)

×	6	2	4	8
6	6	2	4	8
2	2	4	8	6
4	4	8	6	2
8	8	6	2	4

(c)

Figure 33

The associative property is often difficult to justify. Some students just prove it for a particular combination of elements like 3,7 and 9 in A by showing that $3 \times (9 \times 7) = (3 \times 9) \times 7$, but this is not enough. It must be proved to be true for all combinations of three elements in the set, i.e., for $4 \times 4 \times 4 = 4^3 = 64$ triples in all (although some of these triples are trivial).

A general proof is needed or in this case an argument that can be extended to apply to all triples in general.

All members of the set containing 3 can be expressed as $3 + 10n$.

$$\text{So } (3 \times 9) \times 7 = (3 + 10n) \times (9 + 10p) \times (7 + 10q) \quad n, p, q \text{ are integers}$$
$$= (27 + 90n + 30p + 100pn) \times (7 + 10q) \tag{1}$$
$$= (3 \times 9) \times 7 + \text{multiples of } 10$$

Similarly $3 \times (9 \times 7) = 3 \times (9 \times 7) + \text{multiples}$ of $10 = 3 \times 9 \times 7$ (mod 10). Since $3 \times (9 \times 7) = (3 \times 9) \times 7$ for integers the same result holds for integers reduced modulo 10, because all the terms neglected from (1) are multiples of 10. This argument can be applied to every single combination of three elements from set A and set B so in both tables the operation is associative.

The identity in A is 1 and in B is 6.

In A, 1 and 9 are self-inverse and 3 and 7 form an inverse pair.

In B, 6 and 4 are self-inverse and 2 and 8 form an inverse pair.

Both sets A and B form groups, the four conditions being satisfied. If 5 were a member of set B its combination with each element would produce 0. The set would not be closed. Even if 0 were included it would not have an inverse since there is no number which combines with 0 to give the identity 6.

If 5 were included in A, its combination with any of 1, 3, 7 or 9 would produce 5, and 5 would have no inverse. In both cases 5 behaves like a zero element.

In Figure 33(a) the table has been constructed in the order 1,3,9,7 to emphasize the cyclic property. In Figure 33(c) the order 6,2,4,8 produces the cyclic group of order 4 so the 2 groups have the same structure, i.e., they are isomorphic.

Example 23

Derive the group generated by the functions $f(x) = \dfrac{1}{x}$ and $g(x) = 1 - x$ under the combination of functions law where fg means g followed by f assuming that the combination of functions is associative.

First notice that $f^2 = ff(x) = x$ and call this the identity $I(x) = I$. Complete the combination table for the three elements I, f, g (see Figure 34(a)). Two new elements are generated, fg and gf, which are not equal.

$$fg(x) = \frac{1}{1-x} \quad \text{and} \quad gf(x) = 1 - \frac{1}{x} = \frac{x-1}{x}$$

These can be added to the table. Notice that $(gf)f = gf^2 = g$ since

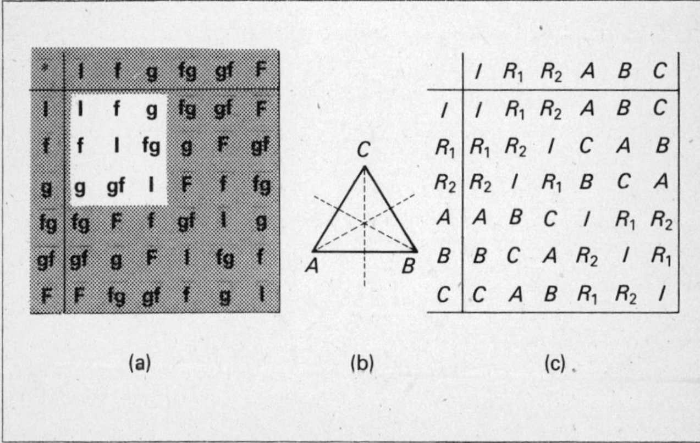

Table (a):

*	I	f	g	fg	gf	F
I	I	f	g	fg	gf	F
f	f	I	fg	g	F	gf
g	g	gf	I	F	f	fg
fg	fg	F	f	gf	I	g
gf	gf	g	F	I	fg	f
F	F	fg	gf	f	g	I

Table (c):

	I	R_1	R_2	A	B	C
I	I	R_1	R_2	A	B	C
R_1	R_1	R_2	I	C	A	B
R_2	R_2	I	R_1	B	C	A
A	A	B	C	I	R_1	R_2
B	B	C	A	R_2	I	R_1
C	C	A	B	R_1	R_2	I

(a)   (b)   (c)

Figure 34

$f^2 = I$ and $f(fg) = f^2 g = g$ since $g^2 = 1 - (1 - x) = x = I$.

$$g(gf) = g^2 f = f \quad \text{and} \quad (fg)g = fg^2 = f$$

These results are easily deduced knowing the associative law holds.

$$fgf(x) = \cfrac{1}{\cfrac{x-1}{x}} = \frac{x}{x-1} : \text{a new function};$$

$$gfg = 1 - \frac{1}{1-x} = \frac{1-x-1}{1-x} = \frac{-x}{1-x} = \frac{x}{x-1}$$

So $fgf = gfg = F$ and the rest of the group table can be derived algebraically.

For instance $\quad (gf)(gf) = g(fgf) = g(gfg) = g^2 fg = fg$
and $\quad\quad\quad\quad (fg)(fg) = f(gfg) = f(fgf) = f^2 gf = gf$

The last row and column for F is now evaluated.

$fF = ffgf = gf; \quad gF = ggfg = fg; \quad Ffg = fgffg = fgg = f$
$fgF = fggfg = g; \quad gfF = gffgf = f; \quad Fgf = gfggf = g$
$FF = fgffgf = I$ or $gfggfg = I$

The last element F completes the set which forms a group of order 6. The group is not commutative as $fg \neq gf$.

Example 24
Form the group table for the symmetry operations of the square.

I is the identity, Q a quarter-turn (rotation 90°), H a half-turn and T a three quarters-turn, all about the centre of the square. These form a commutative cyclic sub-group so the top left-hand corner is easily done. X and Y are reflexions in the horizontal and vertical lines of symmetry and M and N are reflexions in the diagonal lines of symmetry (see Figure 35).

*	I	Q	H	T	X	Y	M	N
I	I	Q	H	T	X	Y	M	N
Q	Q	H	T	I	M	N	Y	X
H	H	T	I	Q	Y	X	N	M
T	T	I	Q	H	N	M	X	Y
X	X	N	Y	M	I	H	T	Q
Y	Y	M	X	N	H	I	Q	T
M	M	X	N	Y	Q	T	I	H
N	N	Y	M	X	T	Q	H	I

The product of a direct and an

opposite isometry is opposite, so
the top right and bottom left
hand corners will consist of X, Y,
M and N while the bottom right
will consist of I, Q, H and T.

Since two reflexions give a
rotation of double the angle
between the mirrors $XY = YX =$
H; also $HX = XH = Y$ and
$HY = YH = X$ to give a sub-
group (I,H,X,Y). This works for

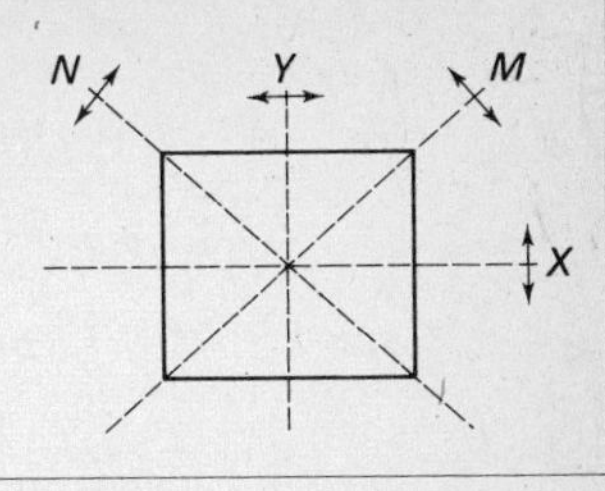

Figure 35

(I,H,M,N) since M and N are reflexions in mirrors at right angles
to each other. MX means X first then M; two reflexions in mirrors
at $45° =$ rotation of $90° = Q$. XM will be the inverse of MX (since
$XMMX = I$) so $MX = T$ and we now have the bottom right hand
corner knowing that there is one of each element in each row and
column.
QX is X first then Q which gives $M \cdot XQ = N$.
$QM = Y$ and $MQ = X$ and hence the complete table is obtained.

Chapter 13

Probability

Multiple choice questions (Type A) Select the correct answer

Example 1

The probability of throwing 3 consecutive heads with three throws of a coin is

A $\frac{1}{2}+\frac{1}{2}+\frac{1}{2}$ **B** $\frac{1}{6}$ **C** $\frac{1}{2}\times\frac{1}{2}\times\frac{1}{2}$ **D** $\frac{1}{2}$ **E** $\frac{1}{4}$

There are 4 different results with 3 throws of a coin: 3 heads; 3 tails; 2 heads and a tail; 2 tails and a head; but each of the last two can occur in 3 ways making 8 possible outcomes in all. The result, 3 heads, is only one result from these 8:

$$\text{HHH, HHT, HTH, THH, HTT, THT, TTH, TTT}$$

The probability is one in eight. **ANSWER C**

Example 2

A set T of numbers has mean 5 and standard deviation 3. If each number in T is doubled, T has mean m and standard deviation s where

A $m = 5, s = 3$ **B** $m = 5, s = 6$ **C** $m = 10, s = 3$
D $m = 10, s = 6$ **E** $m = 10, s = 9$

If $T = x_1, x_2, \ldots, x_n$ then $m = \dfrac{\Sigma x}{n}$ and $s^2 = \dfrac{1}{n}\Sigma (x-m)^2$

For the doubled set $m = \dfrac{\Sigma 2x}{n} = \dfrac{2\Sigma x}{n} = 2 \times 5 = 10$ and

$$s^2 = \frac{1}{n}\Sigma (2x-2m)^2 = \frac{1}{n}\Sigma 2^2(x-m)^2 = \frac{4}{n}\Sigma (x-m)^2 = 4 \times 9 = 36$$

$s = 6$ and $m = 10$. **ANSWER D**

Example 3

The probability of drawing an ace from a pack of cards is $\frac{1}{13}$ and the probability of drawing a heart is $\frac{1}{4}$. The probability of drawing the ace of hearts is

A $\frac{1}{4}+\frac{1}{13}$ **B** $\frac{1}{4}\times\frac{1}{13}$ **C** $\frac{1}{4}-\frac{1}{13}$ **D** $\frac{1}{4}\div\frac{1}{13}$ **E** $\frac{4}{13}$

The ace of hearts is one card out of a pack of 52, so the probability of drawing it is $\frac{1}{52}$.   ANSWER B

In this case we need the probability of both an ace and a heart, i.e., both events happening;
Probability of A and $B = p(A) \times p(B)$ if the events are independent. In this case it is easy to see that the other answers are mostly ridiculous.

Example 4

Three discs (identical except for colour) are drawn from a bag containing 2 red, 3 blue and 4 green discs, without replacement. The probability of drawing 1 red and 2 blue discs in that order is

A $\frac{2}{9} \times \frac{3}{8} \times \frac{3}{7}$ **B** $\frac{2}{9} + \frac{3}{8} + \frac{2}{7}$ **C** $\frac{2}{9} \times \frac{3}{8} \times \frac{2}{7}$ **D** $\frac{1}{2} \times \frac{1}{3} \times \frac{1}{3}$ **E** $\frac{2}{5} \times \frac{3}{4} \times \frac{2}{3}$

The probability that the first disc drawn is red is $\frac{2}{9}$.
This leaves 8 discs, 3 of which are blue, so the probability of the second being blue is $\frac{3}{8}$ which leaves 2 blue out of 7 and the probability of the third being blue is $\frac{2}{7}$.
The probability of all three events is $\frac{2}{9} \times \frac{3}{8} \times \frac{2}{7}$.   ANSWER C

Example 5

In my tutor group of 15, 5 study Mathematics, 10 English and 2 study both. If I pick one at random the probability that he or she studies neither Mathematics nor English is

A 0 **B** $1 - \frac{1}{3} \times \frac{2}{3}$ **C** $\frac{2}{3} \times \frac{1}{3}$ **D** $\frac{2}{15}$ **E** none of these

---

Out of 15, 3 study Mathematics and not English and 8 study English and not Mathematics. With 2 studying both, this accounts for 13, so the 2 left study neither, i.e., 2 out of 15.   ANSWER D

Example 6

The probability that it will rain on any particular day is $\frac{3}{5}$. The probability that it will rain on three out of the next five days is

A 1 **B** $3 \times \frac{3}{5}$ **C** $\frac{3}{5} \times \frac{3}{5} \times \frac{3}{5}$ **D** $(\frac{3}{5})^3 \times (\frac{2}{5})^2$ **E** $10 \times (\frac{3}{5})^3 \times (\frac{2}{5})^2$

The question implies that it rains on three days and does not rain on the other two. The probabilities are therefore $\frac{3}{5} \times \frac{3}{5} \times \frac{3}{5} \times \frac{2}{5} \times \frac{2}{5}$

for three days rain and two without but these can be arranged in $_5C_3$ ways. Total probability is $_5C_3 \times (\frac{3}{5})^3 \times (\frac{2}{5})^2 = 10 \times (\frac{3}{5})^3 \times (\frac{2}{5})^2$.

ANSWER E

Example 7

Ten coins are thrown. The probability of 5 heads appearing is

A $\frac{1}{2}$ **B** $(\frac{1}{2})^5 \times (\frac{1}{2})^5$ **C** $_{10}C_5 \times (\frac{1}{2})^{10}$ **D** $5 \times (\frac{1}{2})^5$
E $5 \times (\frac{1}{2})^5 \times (\frac{1}{2})^5$

Statement C is correct. The result of 5 heads implies 5 tails. These can be arranged in $_{10}C_5$ ways.

ANSWER C

Multiple choice questions (Type B) Answer according to the table

A	B	C	D	E
1, 2, 3	1, 3	2, 3	2	3
correct	only	only	only	only

Example 8

$p(A) = 0.4$, $p(B) = 0.3$, $p(A \text{ or } B) = 0.6$

1 $p(A \cap B) = 0.7$ **2** $p(A \cap \bar{B}) = 0.3$ **3** $p(A|B) = \frac{1}{3}$

$$p(A) + p(B) = p(A \cap B) + p(A \cup B)$$

$$0.4 \quad 0.3 \qquad\qquad 0.6$$

where $P(A \cap B)$ is probability of A and B and $p(A \cup B)$ is probability of A or B or both.

$\Rightarrow p(A \cap B) = 0.1$ so statement 1 is false. This result is similar to the result for the intersection and union of sets.

$p(A \cap \bar{B}) = p(A) - p(A \cap B) = 0.4 - 0.1 = 0.3$ Statement 2 is true

$$p(A|B) = \frac{p(A \cap B)}{p(B)} = \frac{0.1}{0.3} = \frac{1}{3}$$ Statement 3 is true

ANSWER C

Example 9

In throwing 2 dice, S is the event of a total score of 7 and E is the event 'only one of the numbers is even'.

1 $p(S) = \frac{7}{36}$ **2** $p(E) = \frac{1}{2}$ **3** S and E are independent

There are 36 possible results when 2 dice are thrown. Six of these (1,6), (6,1), (2,5), (5,2), (3,4), (4,3) give a total of 7 so
$$p(S) = \tfrac{6}{36} = \tfrac{1}{6} \qquad \text{Statement 1 is wrong}$$
Statement 2 is true, there being 18 pairs in which only 1 number is even.
$p(E|S) = \tfrac{6}{6} = 1$; $P(E|\bar{S}) = \tfrac{12}{30} = \tfrac{2}{5}$ so E and S are not independent.
$p(S|E) = \tfrac{6}{18} = \tfrac{1}{3}$; $p(S|\bar{E}) = 0$ which confirms this.
Statement 2 only is true. $\qquad\qquad$ ANSWER D

Example 10

In drawing a card from a pack of 52 playing cards, the probability of drawing

1 a red card or an ace is $\tfrac{7}{13}$
2 a red card or a black card is $\tfrac{1}{2}$ $\quad$ **3** a jack is $\tfrac{1}{13}$

There are 26 red cards (including 2 red aces) and a further 2 black aces. The probability of a red card or an ace is $\tfrac{28}{52} = \tfrac{7}{13}$.
The probability of a red card is $\tfrac{1}{2}$ and the probability of a black card is $\tfrac{1}{2}$, but whichever card is drawn must be red or black, so the probability of a red or black card will be 1.
There are 4 jacks in the pack so the probability of a jack is $\tfrac{4}{52} = \tfrac{1}{13}$.
Statements 1 and 3 are true. $\qquad\qquad$ ANSWER B

Example 11

The probability of success in a single trial is a. The probability of failure is b. In a series of n such independent trials the distribution of the number of successes has mean m and standard deviation s.

1 $a+b = 1$ $\quad$ **2** $m = na$ $\quad$ **3** $s^2 = nab$

Each trial must result in either success or failure, so statement 1 is true. The probability generator for n trials is $G(t) = (b+at)^n$ where the coefficients of the powers of t in the expansion give the probabilities of the number of successes.

$$\begin{aligned}
m = G'(1) &= na(b+a)^{n-1} = na \\
s^2 = G'(1)+G''(1)-m^2 &= na+n(n-1)a^2(b+a)^{n-2} - m^2 \\
&= na+n(n-1)a^2 - n^2a^2 \\
&= na+n^2a^2 - na^2 - n^2a^2 \\
&= na-na^2 = na(1-a) = nab
\end{aligned}$$

Statements 1, 2 and 3 are all true. $\qquad\qquad$ ANSWER A

> **Example 12**
>
> If 6 dice are thrown together
> **1** the expected number of sixes is 1
> **2** the probability of one six is $(\frac{1}{6}) \times (\frac{5}{6})^6$
> **3** the probability of no sixes is $(\frac{5}{6})^6$

The expected number of sixes is the mean number of sixes and 6 dice thrown together can be regarded as 1 die thrown 6 times. From Example 11, the mean number is $6 \times (\frac{1}{6}) = 1$.
Statement 1 is true.

If one six has occurred so have 5 non-sixes giving a probability of $(\frac{1}{6}) \times (\frac{5}{6})^5$, but this can occur in any one of six ways (the six can occur on any one of the 6 dice) so the total probability is $6 \times (\frac{1}{6}) \times (\frac{5}{6})^5$. Statement 2 is wrong.
No sixes imply all non-sixes, i.e., $(\frac{5}{6})^6$.
Statement 3 is correct. **ANSWER B**

> **Example 13**
>
> Five letters are chosen from the word FAVOURITES. The probability that
>
> **1** they are all vowels is $\frac{1}{2}$
>
> **2** they are all consonants is $\dfrac{1}{{}_{10}C_5}$
>
> **3** there are more vowels than consonants is $\frac{1}{2}$

The probability of all vowels or all consonants must be the same as there are 5 of each. There is only one way of choosing 5 vowels out of ${}_{10}C_5$ ways of choosing 5 letters so statement 1 is wrong and statement 2 is correct. In selecting 5 letters all the letters are different so each selection including 1 vowel contains 4 consonants, and 2 vowels gives 3 consonants, etc. Half the selections contain more vowels than consonants and half do not; 3 is correct. **ANSWER C**

Multiple choice questions (Type C) For each question two statements are given. Answer

A if 1 implies 2 and 2 implies 1
B if 2 implies 1 but 1 does not imply 2
C if 1 implies 2 but 2 does not imply 1
D if 1 denies 2 and 2 denies 1
E if none of these relations holds

> **Example 14**
> **1** The probability that a blue-eyed person is colour-blind is $\frac{1}{5}$.
> **2** The probability that a colour blind person is blue-eyed is $\frac{1}{5}$.

This question deals with the relation between the conditional probabilities $p(A|B)$, the probability of A given B, and $p(B|A)$, the probability of B given A. In general these are not equal.

If $p(B) = b$ and $p(A|B) = a$ then $p(A \cap B) = p(B) \times p(A|B) = ba$.

If in addition $p(A|\bar{B}) = a_1$, then $p(A \cap \bar{B}) = p(\bar{B} \cap A) = (1-b)a_1$

$$p(A) = p(A \cap B) + p(A \cap \bar{B}) = ba + (1-b)a_1$$

$$p(B|A) = \frac{p(B \cap A)}{p(A)} = \frac{ba}{ba + (1-b)a_1} \text{ which is not equal to } p(A|B) = a.$$

If A and B are independent events then $p(A|B) = p(A|\bar{B}) \Rightarrow a = a_1$

and $p(B|A) = \dfrac{ba}{ba + (1-b)a} = \dfrac{ba}{a} = b = p(B) \doteq p(B|\bar{A})$.

But in the question, statement 1 implies statement 2 would require a to be equal to b and in general this is not true. However, it could be true, so statement 1 does not deny statement 2.   ANSWER E

> **Example 15**
> **1** The probability of a person being left-handed is $\frac{1}{5}$.
> **2** In a random sample of five people, one person only is found to be left-handed.

If statement 1 is true, then the probability of finding statement 2 true is

$$5 \times \left(\frac{1}{5}\right) \times \left(\frac{4}{5}\right)^4 = \frac{4^4}{5^4} = \frac{256}{625} \simeq 0.41$$

If statement 2 is true, there is no guarantee that statement 1 is true. However statements 1 and 2 do not deny each other. ANSWER E

In fact if statement 1 is true we can say that the larger the sample taken then the more closely the fraction of left-handed people approaches $\frac{1}{5}$.

If statement 2 is true we have considered only a very small sample and even if we took many samples of only 5 people we would still not be convinced that statement 1 is true.

> **Example 16**
> **1** $p(B|A) = p(B|\bar{A})$ **2** A and B are independent.

Statement 1 is the definition of statistical independence. If $p(B|\bar{A}) = p(B|A)$ then $p(B)$ is equal to both of these and the probability of B does not depend upon whether A has happened or not. Statement 1 implies statement 2 and vice versa. **ANSWER A**

> ### Example 17
> **1** A and B are exhaustive events **2** $p(A \cap B) = 0$

If A and B are exhaustive then they exhaust all possibilities. This means that the probability of A or B (or both) is equal to 1.
$p(A \text{ or } B) = p(A \cup B) = p(A) + p(B) - p(A \cap B)$
$p(A \cap B) = 0$ does not guarantee that $p(A \cup B) = 1$. It could be however so statement 2 does not imply statement 1, nor deny it.

In the throw of a single die, if A is throwing an even number and B is throwing a prime (counting 1 as prime) then $p(A) = \frac{1}{2}$ and $p(B) = \frac{2}{3}$, $p(A \text{ or } B) = 1$ but $p(A \cap B) = \frac{1}{6}$ so statement 1 does not imply statement 2. Statement 1 does not deny statement 2 either since in the toss of a coin if $p(A) = p(\text{tail}) = \frac{1}{2}$ and $p(B) = p(\text{head}) = \frac{1}{2}$, $p(A \text{ or } B) = 1$ and $p(A \cap B) = 0$. **ANSWER E**

> ### Example 18
> **1** A and B are exclusive events **2** $p(A \cup B) = 1$

In these last two questions, the conditions for exhaustive and exclusive events have been interchanged, i.e.
'exhaustive' $\Leftrightarrow p(A \cup B) = 1$ and 'exclusive' $\Leftrightarrow p(A \cap B) = 0$
A and B exclusive does not confirm or deny statement 2, and similarly statement 2 does not confirm or deny statement 1. **ANSWER E**

> ### Example 19
> **1** A and B are independent **2** $p(A \cap B) = p(A) \times p(B)$

$p(A \cap B) = p(A) \times p(B|A)$
A and B independent $\Rightarrow p(B|A) = p(B|\bar{A}) = p(B) \Rightarrow p(A \cap B) = p(A) \times p(B)$.
Statement 1 implies statement 2.

$$p(A \cap B) = p(A) \times p(B) \Rightarrow p(B) = \frac{p(A \cap B)}{p(A)} = p(B|A)$$

$$p(B|\bar{A}) = \frac{p(B \cap \bar{A})}{p(\bar{A})} = \frac{p(B) - p(A \cap B)}{1 - p(A)} = \frac{p(B) - p(A) \times p(B)}{1 - p(A)} = p(B)$$

$= p(B|A)$ so B is independent of A.
Statement 1 implies statement 2 and vice versa.   **ANSWER A**

Multiple choice questions (Type D) Each question consists of a problem followed by four pieces of information. Decide whether the problem can be solved with one of the four pieces of information omitted and answer

A if 1 could be omitted
B if 2 could be omitted
C if 3 could be omitted
D if 4 could be omitted
E if none can be omitted

Example 20

Only a certain number, N, of a pack of 52 playing cards are shuffled and then cut 4 times. Find the probability of cutting a club, a diamond, a heart and a spade, in that order. Of the pack of N cards:

1 $\frac{1}{5}$ are hearts   **2** $\frac{1}{2}$ are red   **3** $\frac{1}{4}$ are clubs
4 5 cards are spades

Values 1 and 2 imply $\frac{1}{5}$ of N are hearts and therefore $\frac{3}{10}$ of N are diamonds. If $\frac{1}{4}$ are clubs, then $\frac{1}{4}$ are spades and the probability of drawing club, diamond, heart, spade is $\frac{1}{4} \times \frac{3}{10} \times \frac{1}{5} \times \frac{1}{4} = \frac{3}{800}$. Value 4 is not needed.   **ANSWER D**

Example 21

In a group of girls and boys, find the probability of choosing at random a left-handed girl.

1 number of boys $= 10$ **2** number of right-handed boys $= 8$
3 number of girls $= 12$ **4** number of left-handed people $= 5$

Values 1 and 3 give the total of the group as 22. Now we need to know the number of left-handed girls. Value 2 gives the number of left-handed boys as $10 - 8 = 2$. We then need value 4 to give us the number of left-handed girls to be $5 - 2 = 3$.

The probability of a left-handed girl is $\frac{3}{22}$. All four values are needed.   **ANSWER E**

> **Example 22**
> Find how many staff, on average, arrive late at Woking College.
>
> **1** The staff arrival times form a normal distribution.
> **2** There are 70 members of staff.
> **3** The mean arrival time is 8.50 a.m.
> **4** The arrival times have a standard deviation of 5 minutes.

The mean and standard deviation specify a normal distribution, so from values 1, 3 and 4, the fraction of staff arriving after any specified time can be determined. Value 2 (70 staff) enables the actual number of staff (not the fraction) to be calculated, but we must also know that lessons begin at 9 a.m. The question needs all four pieces of information and more. ANSWER E

Short questions

> **Example 23**
> Two unbiased dice are thrown and the sum, S, is recorded on each throw. Find the mean and standard deviation of S.

The values for S range from 2 to 12 with the following probabilities

S	2	3	4	5	6	7	8	9	10	11	12
Probability	$\frac{1}{36}$	$\frac{2}{36}$	$\frac{3}{36}$	$\frac{4}{36}$	$\frac{5}{36}$	$\frac{6}{36}$	$\frac{5}{36}$	$\frac{4}{36}$	$\frac{3}{36}$	$\frac{2}{36}$	$\frac{1}{36}$

The mean m of S is $m = \Sigma S \times \text{Prob}(S) = 2 \times \frac{1}{36} + 3 \times \frac{2}{36} + 4 \times \frac{3}{36} + \ldots$
$m = (2 + 6 + 12 + 20 + 30 + 42 + 40 + 36 + 30 + 22 + 12) \div 36$
$= 252 \div 36 = 7$
This result is to be expected as the probability distribution is symmetrical.

The standard deviation is given by $\text{S.D.}^2 = \Sigma (S - m)^2 pr(S)$

$$\text{S.D.}^2 = (2-7)^2 \times \tfrac{1}{36} + (3-7)^2 \times \tfrac{2}{36} + (4-7)^2 \times \tfrac{3}{36} + \ldots$$

$$= 2 \times 5^2 \times \tfrac{1}{36} + 2 \times 4^2 \times \tfrac{2}{36} + 2 \times 3^2 \times \tfrac{3}{36} + 2 \times 2^2 \times \tfrac{4}{36}$$

$$+ 2 \times 1^2 \times \tfrac{5}{36} + 0$$

$$= \tfrac{1}{18}(25 + 32 + 27 + 16 + 5) = \tfrac{105}{18} = 5\tfrac{5}{6} = 5.8\dot{3}$$

$$\text{S.D.} = \sqrt{5.8\dot{3}} \simeq 2.42$$

There are short cuts to calculations like this but in this case where the distribution is symmetrical (the mean being 7) it is just as quick to use the basic definitions of mean and S.D.

> **Example 24**
> On my journey to work 20 per cent of the cars I see contain more than one person. On a particular morning, what is the probability that three of the first five cars I see contain more than one person?

The Binomial Probability Generating Function is $(\frac{4}{5}+\frac{1}{5}t)^5$
For three cars the coefficient of t^3 is needed, i.e.
$p(3 \text{ cars with more than 1 person}) = 10(\frac{4}{5})^2 \times (\frac{1}{5})^3$

$$= \frac{160}{5^5} = \frac{32}{5^4} \simeq 0.0512$$

The probability of seeing 3 or more cars with more than 1 person is

$$p(3)+p(4)+p(5) = 10(\frac{4}{5})^2(\frac{1}{5})^3 + 5(\frac{4}{5})(\frac{1}{5})^4 + 1(\frac{4}{5})^5$$

$$= \frac{160+20+1}{5^5} = \frac{181}{5^5} \simeq 0.058$$

> **Example 25**
> $p(X) = \frac{1}{5}$ $p(Y|X) = \frac{1}{4}$ $p(Y|\bar{X}) = \frac{1}{3}$ find $p(X|Y)$

$$p(X \cap Y) = p(X) \times p(Y|X) = \frac{1}{5} \times \frac{1}{4} = \frac{1}{20}; \; p(\bar{X} \cap Y)$$
$$= p(\bar{X}) \times p(Y|\bar{X}) = \frac{4}{5} \times \frac{1}{3} = \frac{4}{15}$$

$$p(Y) = p(X \cap Y) + p(\bar{X} \cap Y) = \frac{1}{20} + \frac{4}{15} = \frac{3+16}{60} = \frac{19}{60}$$

$$p(X|Y) = p\frac{(X \cap Y)}{p(Y)} = \frac{\frac{1}{20}}{\frac{19}{60}} = \frac{3.}{19}$$

> **Example 26**
> Given $p(A) = \frac{2}{5}$, $p(B) = \frac{1}{4}$ and $p(A \cap B) = \frac{3}{20}$, find out whether A and B are statistically independent.

We need to find the conditional probabilities $p(A|B)$ and $p(A|\bar{B})$ or $p(B|A)$ and $p(B|\bar{A})$.

Method 1 Tree diagram using the notation in the diagram (see Figure 36(a))
$$p(A) \times p(B|A) = p(A \cap B) \Rightarrow \frac{2}{5} \times b_1 = \frac{3}{20} \Rightarrow b_1 = p(B|A) = \frac{3}{8}$$
$$p(B) = p(A \cap B) + p(\bar{A} \cap B) \Rightarrow \frac{1}{4} - \frac{3}{20} + \frac{3}{5}b_2 \Rightarrow \frac{3}{5}b_2 = \frac{1}{10} \Rightarrow b_2$$
$$= p(B|\bar{A}) = \frac{1}{6}$$
$p(B|A) \neq p(B|\bar{A}) \Rightarrow A$ and B are not independent. They are dependent.

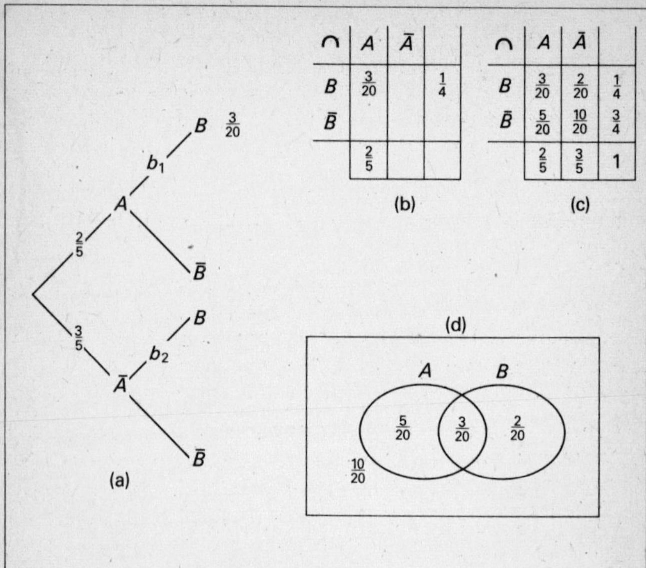

Figure 36

Method 2 Contingency table. In Figure 36(b) the information given is recorded. On the top line, since $p(B) = \frac{1}{4} = p(A \cap B) + p(\bar{A} \cap B)$, the missing value $p(B \cap \bar{A})$ is calculated as $\frac{1}{4} - \frac{3}{20} = \frac{2}{20}$. Similarly the rest of the table is calculated (see Figure 36(c)).

$$p(B|A) = \frac{\frac{3}{20}}{\frac{2}{5}} = \frac{3}{8} \text{ (1st column)} \quad p(B|\bar{A}) = \frac{\frac{2}{20}}{\frac{3}{5}} = \frac{1}{6} \text{ (2nd column)}$$

Method 3 Venn diagram. Overlap $\equiv p(A \cap B) = \frac{3}{20} \Rightarrow$ LHS $= p(A \cap \bar{B}) = \frac{2}{5} - \frac{3}{20} = \frac{5}{20}$ and RHS $= p(B \cap \bar{A}) = \frac{1}{4} - \frac{3}{20} = \frac{2}{20} \Rightarrow$ Outside $= p(\bar{A} \cap \bar{B}) = \frac{10}{20}$ (see Figure 36(d))

$$p(B|A) = \frac{\frac{3}{20}}{\frac{8}{20}} = \frac{3}{8} \text{ and } p(B|\bar{A}) = \frac{\frac{2}{20}}{\frac{12}{20}} = \frac{1}{6} \Rightarrow A \text{ and } B \text{ are dependent.}$$

Example 27

I belong to a 100 Club, where 100 members pay in £1 per month. There are prizes of £24 and £12 each month except in June when there is one prize of £100 and in December when there is one prize of £200. What are my chances of breaking even (winning at least £12 during the year)?

The probability of winning £200 is 1 in 100, i.e., 0.01.
The probability of winning £100 is 1 in 100, i.e., 0.01. In each of the other months the probability of winning £24 is 1 in 100 and the probability of winning £12 is 1 in 99. This happens 10 times during the year, so the probability of winning £12 or £24 during the year is

$$10\left(\frac{1}{100}+\frac{1}{99}\right)=\frac{1990}{9900}.$$

Total probability of breaking even (at least I may win more than 1 prize)

$$=\frac{1990}{9900}+\frac{1}{100}+\frac{1}{100}=\frac{2188}{9900}=0.221$$

I have almost a 1 in 4 chance of breaking even. How does the Club make money? With 100 members having a 1 in 4 chance of breaking even is there a profit?

The answer is that the probability of each member winning is not independent from each other. Not every member can win. In fact there are only 22 prizes and yet 100 members.

The 100 Club is superficially attractive to each member and yet it pays out £660 in prizes and makes £540 profit per year.

Example 28
If the probability of passing each driving test is $\frac{1}{3}$, find the mean number of tests each person takes.

The probability of taking 1 test is $\frac{1}{3}$.
The probability of taking 2 tests is $\frac{2}{3}\times\frac{1}{3}$

(fail the 1st, pass the 2nd)

The probability of taking 3 tests is $(\frac{2}{3})^2\times\frac{1}{3}$

(fail 1st and 2nd, pass 3rd)

Probability generator $G(t)=\frac{1}{3}t+\frac{2}{3}\times\frac{1}{3}t^2+(\frac{2}{3})^2\times\frac{1}{3}t^3+\ldots=\dfrac{\frac{1}{3}t}{1-\frac{2}{3}t}$

(Summing this as a converging infinite G.P.)

Where the probability of taking n tests is the coefficient of t^n, the mean number of tests is given by

$$1\times\tfrac{1}{3}+2\times\tfrac{2}{3}\times\tfrac{1}{3}+3\times(\tfrac{2}{3})^2\times\tfrac{1}{3}+\ldots=G'(1)$$

$$G'(t)=\frac{d}{dt}\left(\frac{t}{3-2t}\right)=\frac{(3-2t)1-t(-2)}{(3-2t)^2}=\frac{3}{(3-2t)^2}\Rightarrow G'(1)=3$$

Mean number of tests taken is 3.

> **Example 29**
>
> In our Sixth Form College, where $\frac{3}{5}$ of the students are female, each of 44 tutor groups elects a representative. From these 44 representatives an executive committee of 6 are elected. What is the likely composition of the executive committee (a) if girls always vote for girls and boys always vote for boys (b) if each votes for the opposite sex.

(a) The probability of selecting male and female representatives for the tutor groups will be $\frac{2}{5}$ and $\frac{3}{5}$.

The probability of a male or female representative being elected onto the executive committee will be $\frac{2}{5}$ and $\frac{3}{5}$.

Therefore the probability of a male representative being elected is $\frac{2}{5} \times \frac{2}{5} = \frac{4}{25}$ and the probability of a female representative being elected is $\frac{3}{5} \times \frac{3}{5} = \frac{9}{25}$.

The ratio will be $9:4 \equiv$ female:male, so the executive committee will have 4 girls and 2 boys.

Or could we argue, that in the tutor groups the male:female ratio is $2:3$ so the candidates proposed will be in that ratio. Thus after voting (in that ratio) the tutor group representatives will be in the ratio $4:9$ and after voting for the executive committee, the male:female ratio will be $8:27$.

(b) The probability of selecting male:female representatives is $\frac{3}{5}:\frac{2}{5}$. The probability of a male representative being elected onto the executive committee is $\frac{3}{5} \times \frac{3}{5}$ and female $\frac{2}{5} \times \frac{2}{5}$, i.e., $9:4 \equiv$ male:female.

Or is it that the ratio of male:female proposed is $2:3$ so the ratio of tutor group representatives is $\frac{2}{5} \times \frac{3}{5}:\frac{3}{5} \times \frac{2}{5}$, i.e., $1:1$.

So the ratio of male:female executives is $\frac{3}{5}:\frac{2}{5}$, i.e., $3:2$.

> **Example 30**
>
> Find the number of ways of choosing a mixed hockey team of 6 men and 5 women players (if the goalkeeper must be a man) from 9 men and 9 women.

The goalkeeper can be chosen in 9 ways. When that is done the number of ways of choosing 5 men from 8 is $_8C_5 = \dfrac{8 \cdot 7 \cdot 6}{1 \cdot 2 \cdot 3} = 56$.

Number of ways of choosing 5 women from 9 is $_9C_5 = \dfrac{9 \cdot 8 \cdot 7}{1 \cdot 2 \cdot 3} = 84$.

Total number of ways of choosing a team is $56 \times 84 \times 9 = 42\,336$.

Long questions

> **Example 31**
> At Esher College, 2 members of staff out of 50 have arrived by 8.30 a.m. and 15 by 8.40 a.m. Assuming that the arrival times for members of staff form a normal distribution, calculate the mean arrival time and find out how many staff are late for the first period which starts at 9.05 a.m.

If 8.30 a.m. corresponds to $(\text{mean} - x_1)$ S.D.s.

$\phi(+x_1) = 1 - \frac{2}{50} = 0.96$

(This is the proportion of the population less than x_1 S.D.s to the right of the mean.) $x_1 = 1.751$ S.D.s.

Similarly if 8.40 a.m. corresponds to $(\text{mean} - x_2)$ S.D.s

$$\phi(+x_2) = 1 - \tfrac{15}{50} = 0.7 \Rightarrow x_2 = 0.8418 \text{ S.D.s}$$

$10\,\text{min} \equiv 1.751 - 0.8418\,\text{S.D.s} = 0.9092\,\text{S.D.s} \Rightarrow 1\,\text{S.D.} = 11\,\text{min}$

$0.8418\,\text{S.D.s} \equiv 0.8418 \times 11\,\text{min} = 9.259\,\text{min}$

so mean $= 8\,\text{hr}\ 49.259\,\text{min}$

9.05 a.m. is 15.741 min beyond mean, i.e., $\dfrac{15.741}{9.259}\,\text{S.D.s} = 0.588\,\text{S.D.s}$

$\phi(0.588) = 0.7208$, i.e., 0.2792 of the population still to arrive.

$0.2792 \times 50 = 13.96$ members of staff. 14 are late.

Is this a good probability model for the situation?

> **Example 32**
> 4 balls are drawn from a bag containing 3 blue and 5 red ones, the balls being identical except for colour. Find the probabilities of drawing (i) 4 blues, (ii) 4 reds, (iii) 2 of each colour if (a) each ball is replaced after drawing and (b) the balls are not replaced.

(a) Replacing. If the balls are replaced, then the conditions are the same for each drawing, i.e., $p(\text{blue}) = \frac{3}{8}$, $p(\text{red}) = \frac{5}{8}$.

(i) $p(4 \text{ blues}) = \left(\dfrac{3}{8}\right)^4 = \dfrac{3^4}{8^4} = \dfrac{81}{4096} \simeq 0.02$

(ii) $p(4 \text{ reds}) = \left(\dfrac{5}{8}\right)^4 = \dfrac{5^4}{8^4} = \dfrac{625}{4096} \simeq 0.15$

(iii) 2 reds and 2 blues can occur in $_4C_2 = 6$ different ways.

$p(2 \text{ of each}) = 6 \times \left(\dfrac{3}{8}\right)^2 \times \left(\dfrac{5}{8}\right)^2 = \dfrac{1350}{4096} \simeq 0.33.$

(*b*) Without replacing

(i) $p(4 \text{ blues}) = 0$; it cannot be done, as there are only 3 blue balls.

(ii) $p(4 \text{ reds}) = \frac{5}{8} \times \frac{4}{7} \times \frac{3}{6} \times \frac{2}{5} = \frac{1}{14} \simeq 0.07$

(iii) $p(\text{BBRR}) = \frac{3}{8} \times \frac{2}{7} \times \frac{5}{6} \times \frac{4}{5}$; $p(\text{BRBR}) = \frac{3}{8} \times \frac{5}{7} \times \frac{2}{6} \times \frac{4}{5}$, etc.

Each of the 6 ways of drawing 2 of each will give $\frac{3}{8} \times \frac{2}{7} \times \frac{5}{6} \times \frac{4}{5}$ so
$p(2 \text{ of each}) = 6 \times \frac{3}{8} \times \frac{2}{7} \times \frac{5}{6} \times \frac{4}{5} = \frac{6}{14} = \frac{3}{7} \simeq 0.43$.